Broadening the Scope of Wellbeing Science

Andrew H. Kemp • Darren J. Edwards
Editors

Broadening the Scope of Wellbeing Science

Multidisciplinary and Interdisciplinary Perspectives on Human Flourishing and Wellbeing

palgrave
macmillan

Editors
Andrew H. Kemp
School of Psychology
Swansea University
Swansea, UK

Darren J. Edwards
School of Health and Social Care
Swansea University
Swansea, UK

ISBN 978-3-031-18328-7   ISBN 978-3-031-18329-4 (eBook)
https://doi.org/10.1007/978-3-031-18329-4

This Palgrave Macmillan imprint is published by the registered company Springer Nature Switzerland AG.
The registered company address is: Gewerbestrasse 11, 6330 Cham, Switzerland

Foreword: From Positive Psychology to a Wellbeing Science: Past Successes and Future Promise

I am delighted to recommend and endorse this forward-thinking book, in which experts present introductions to and developments in positive psychology and acceptance and commitment therapy, documenting how this work has evolved and risen to meet several grand challenges, including for example, the unfolding climate catastrophe. The book also includes cutting-edge work that focuses on multidisciplinary and multidimensional levels of exploration from the individual to society, broadly consistent with the Extended Evolutionary Meta-Model (EEMM). My colleagues and I developed this EEMM framework in an attempt to move beyond debate in the field, laying strong foundations for improving wellbeing at multiple levels of scale, irrespective of theoretical orientation. In this forward, I will review some common criticisms that have been directed at the field of positive psychology, which are arguably fundamental to the foundations on which wellbeing science is trying to build. I will also present how I believe the field is responding to these criticisms and I then suggest strategies that—together with insights from the authors of subsequent chapters—will allow positive psychology specifically, and wellbeing science more generally, to become a more coherent, integrated, and effective area of research and practice.

Positive Psychology Critiques

The major critiques of positive psychology are that it overemphasises hedonism, underemphasises context, ignores negative experience, and offers nothing new. I believe positive psychology, as a domain, clearly refutes each of these criticisms.

We Cannot Reduce Positive Psychology to Hedonism

Critics have argued that positive psychology reinforces the idea that the ideal person is one who is a cheerful, outgoing, goal-driven, status-seeking extrovert. We should all strive to be a 'materialistic go-getter' or a 'positive thinker'. It is now clear that this view is more of a stereotype of what people think positive psychology is, than an accurate characterisation of what it does. Substantial research within positive psychology has shown that materialism, by itself, rarely produces wellbeing. Other researches show the importance of long-term meaning in life, contrasted with short-term hedonism. In the present volume, the chapter by Wong and colleagues (Chap. 2) characterises the integration of the dark (negative) and light (positive) sides of life as an emergent field of existential positive psychology, while that by Pritchard Richardson (Chap. 6) argues for the transformative power of connecting psychologically to nature. Nature is not just about feeling good, but also about beauty, appreciation, meaning, and transcendence.

Positive Psychology Is Sensitive to Context

Positive psychology has been accused of emphasising individual characteristics (e.g., lack of optimism) as the cause of struggle, and underemphasising the role of context. Problems are seen as a deficit in the person, rather than as a problem with the environment. For example, if someone experiences burnout at work, then a poorly trained coach might assume that the problem is due to a personal characteristic, such as a 'lack' of resilience. This focus ignores the possibility that the workplace is toxic and needs to be changed. We might view this entire book as a refutation of the context-insensitivity argument, as the book focuses on broadening positive psychology to wider contexts. The chapter by Lambert and colleagues (Chap. 3) in particular describes how positive psychology has begun to address systemic issues in society. Positive psychology has clearly moved beyond context insensitivity, with many publications now focusing on the role of social groups, policy and culture influencing wellbeing. In this regard, the chapter by David Neilson (Chap. 10) argues that aspects of capitalism undermine wellbeing. See the chapter by Webber (Chap. 9) also in regard to a related perspective from economics.

Positive Psychology Does Not Ignore Negative Experience

Critics have suggested that positive psychology is pollyannaish and ignores the negative. Seligman targeted the opposite argument at clinical psychology, arguing that clinical psychology was 'too negative' and ignored the positive. Today, we can say that neither critique is correct. For example, in this book, Paul Wong describes how to flourish even in the presence of deep existential anguish and suffering. Indeed, he argues that suffering is an essential ingredient for flourishing. In clinical psychology, ACT helps people live valued, meaningful lives, even when experiencing distress, hardship, and pain. The chapter by Edwards (Chap. 5) describes how ACT has begun to be applied at higher levels of scale within society, beyond the individual, laying foundations for collective wellbeing based on a 'prosocial' framework which utilises the ACT matrix. The chapter by Gibbs and colleagues (Chap. 4) describes how they have applied positive psychotherapy to people living with acquired brain injury (ABI) in the United Kingdom, while also seeking to overcome entrenched socio-structural barriers to wellbeing through systems-informed thinking and strategic partnerships.

Positive Psychology Offers Something New

A final critique of positive psychology is that it is old wine in new bottles. It is not making any conclusions that were not already being made by clinical and 'negative-psychology'. This critique would make sense if we assume that positive and negative are two sides of a continuum. Then, if you reduce negative experience, you will, by definition, be increasing the positive experience. However, this is clearly not the case as the chapter by Gibbs and colleagues (Chap. 4) has shown for people living with ABI. While ABI is associated with the experience of significant hardship, physical impairment, and psychological distress, there remains tremendous potential for wellbeing. Positive and negative experiences are distinct. One can experience high levels of both positive and negative affects (intense people), low levels of both ('disengaged'), or some combination of both. This means that we can know everything there is to know about people's negative behaviour and experience, and this would not tell us much about their positive experience. An entire lifetime of studying negative processes will not necessarily lead to an understanding of positive processes.

Perhaps the clearest refutation of the 'nothing new' critique is the substantial number of new insights that emerged in recent years, including a focus on strength spotting, creativity, positive affect and broaden-and-build theory, flow, self-compassion, value affirmation, optimism, meaning in life, volunteering, helping behaviour, forgiveness, prosocial emotions and gratitude, hope, and self-transcendence. It seems unlikely that an exclusive focus on fixing pathology would have led to so many insights about wellbeing.

After two decades of research, we can say Martin Selgiman was right. Endless focus on the negative was never going to lead to a deeper understanding of the positive. And critically, Seligman's ideas are now moving beyond individual wellbeing with implications for collective and planetary wellbeing. Moreover, the chapter by Edmonds (Chap. 8) further demonstrates how the ideas emerging from positive psychology are beginning to influence professional practice beyond the clinic, including the field of public health.

The Next Steps in Positive Psychology and Wellbeing Science

Dynamic Linking of Positive and Negative

I've argued that positive and negative processes are distinct and are not just two sides of the same dimension (see also chapter by Wong and colleagues, Chap. 2). However, this does not mean that positive and negative are independent. Positive can influence negative, and vice versa. The time has come to bring the positive and negative together, without seeking to reduce them to one dimension. Instead, we need to examine how they interact in a complex network of relations. For example, research shows that there is only a weak correlation between self-criticism and self-compassionate responding, suggesting that people can engage in *both* processes. Thus, instead of two sides of one continuum, we can think of self-compassion and self-criticism as distinct, dynamic processes that influence each other. Increasing self-compassionate responding may help reduce self-criticism for some. Similarly reducing self-criticism may increase self-compassionate responding. This example illustrates that a network-based or systems-informed approach allows for the positive and negative to coexist in a dynamic system of influence. Interventions to increase self-compassion and reduce self-criticism may be different, and yet the two processes are likely to influence each other.

Evidence-Based Processes Instead of Packages

Interventions to improve wellbeing are often like the black box in which we don't quite know what is going on inside. What is working, what is not working, and for whom at an individual level and beyond?

The solution to this problem is to shift our focus from evidence-based packages to evidence-based processes. A process-based approach can break the dictatorship of trademarked packages that require people to do special training to get certified in a complex package approach. Rather, research can identify evidence-based processes and train people in those, depending on their needs and skills. The chapter by Edwards (Chap. 5) for instance, describes how these ideas are now targeting multiple levels of scale from the individual to the group and interactions between groups.

Process and Dimensions at Multiple Levels of Scale

I will now provide the reader with an intuitive way of understanding evidence-based processes. We hope that this will serve as a road map as you read through the book, and will help you situate the different chapters within a 'process-based map'. The Extended Evolutionary Meta-Model (EEMM) provides a framework for organising and talking about different processes of change (the interested reader can explore the EEMM further at josephciarrochi.com or pbatsupport.com). A process of change is a contextually situated, modifiable behaviour or sequence of behaviours that orients the client towards an adaptive outcome. We use the term 'behaviour' broadly here, as there are three levels that a process of change can be described: the physiological level (e.g., heart rate variability as index of stress), the psychological level (hope, optimism), and the social level (group hope, cultural beliefs) .

We can also think of processes as having different dimensions. These include:

1. The *affective dimension*, which focuses on increasing positive affect, reducing reactivity to negative affect, and increasing specific affective states like contentment, tranquillity, and love.
2. The *cognitive dimension*, which focuses on problem solving, reappraising, developing positive attitudes, optimism, and hope.
3. The *attentional dimension*, which focuses on processes related to mindfulness or awareness of feelings, awareness of the consequences of behaviour, savouring or being fully present to positive experience, flow, noticing nature, attentional control, and distractibility.
4. The *self dimension*, which focuses on processes about the self being worthy of love (self-esteem) and able to accomplish things (self-efficacy). It also includes interventions focused on self-compassion, best possible self, and transcendent senses of self. Finally, it includes strength-spotting exercises that help people to understand their skills and capabilities, to understand 'who they are'.
5. The *motivation dimension*, which focuses on identifying strengths that reflect what one loves and values, identifying autonomous versus controlled forms of motivation, clarifying values, and interventions that teach the value of community, sharing, patience, courage, gratitude, responsibility and the value of character.
6. The *overt behaviour dimension*, which focuses on promoting goal striving persistence, patience, courageous behaviour, and positive habits. It also includes processes focused on promoting goal striving persistence, patience, courageous behaviour, and positive habits.

All six dimensions can be located at different dimensions. For example, we might ask, does heart rate variability (physiological) link to our ability for self-compassion (individual). Finally, we can look at the link between the group-level and individual-level processes. Groups often share culture and can include small communities, workplaces, countries, and even groups of countries (e.g., Western versus Eastern). Groups, like individuals, can vary in affect (high levels of stress and burnout), cognition (beliefs and norms), attention (excessive focus on future success), self ('I'm inferior if I'm poor'), motivation (valuing materialism and hedonism over relationships and nature), and overt behaviour (everybody is commuting and working 80 hours a week). Individuals and subgroups may differ substantially from the dominant culture, as when a group of committed environmental activists push back against a cultural agenda that views nature only as a resource to be exploited. Future research is needed to examine how positive psychological processes at the group level influence individual-level processes, and vice versa.

Conclusion

Wellbeing science is a domain with many mature programmes of research that have expanded how we think about human thriving, and our relationship with others and the natural environment. The critiques of the field have been answered in past research, and are directly addressed in this book. Looking forward, positive psychology needs to continue to focus on evidence-based processes of change, rather than evaluating trademarked positive psychology packages. Complementing this endeavour, contributors to this book demonstrate how the field of wellbeing science is beginning to evolve, and point to a need for a more transdisciplinary science of wellbeing, that is more integrated and moves beyond disciplinary boundaries. Just as the heart and brain are distinct and yet utterly interconnected, so too are the wellbeing of the individual, other people, and nature.

Institute of Positive Psychology and Education Joseph Ciarrochi
Australian Catholic University
Sydney, NSW, Australia

Contents

Notes on Contributors

Sabrina Al Subaiei is a Saudi national graduate psychology student currently completing a master's degree in the USA. She is an alumnus of the United Arab Emirates University (UAEU) and the Sheikh Mohamed bin Zayed Scholars Program at NYU Abu Dhabi (UAE).

Victoria L. Bowers holds a PhD in clinical psychology and an MA in psychology from Saybrook University. She is a registered psychological assistant at City Psychology Inc., in Modesto, California, where she completed her postdoctoral internship and continues to remain on staff. She works from an existential-humanistic perspective and is trained in psychodynamic theory, cognitive behavioural therapy (CBT), and dialectical behavioural therapy (DBT). Bowers has clinical experience working with severe mental illness, trauma, domestic violence, sexual assault, and homelessness, and treats diverse populations including families, couples, children, and groups. Bowers is a member of APA (American Psychological Association) Division 36, The Society for the Psychology of Religion and Spirituality, where she serves on a task force for early-career psychologists. She is currently working towards becoming a licensed clinical psychologist and plans to open a nonprofit agency dedicated to education and research on mental illness. Bowers enjoys advocacy work for underprivileged youth and is the happiest when she is in the service of others.

Joseph Ciarrochi, PhD is a professor at the Institute for Positive Psychology and Education at Australian Catholic University. He has published more than 160 scientific journal articles and many books, including the widely acclaimed *Emotional Intelligence in Everyday Life* and *The*

Weight Escape. His latest book is *What Makes You Stronger: How to Thrive in the Face of Uncertainty Using Acceptance and Commitment Therapy*. Ciarrochi has been honoured with more than four million dollars in research funding. His work has been discussed on TV and radio, and in magazines and newspaper articles. He is ranked in the top 1% of scientists in the world across all disciplines.

Nadia S. Corral-Frías, PhD is an associate professor at the University of Sonora, Mexico. She completed her undergraduate degree with honours in psychology at Instituto de Estudios Superiores del Occidente in Guadalajara, Mexico. She went on to the University of Arizona and earned a degree in neuroscience and a postdoctoral fellowship at Washington University in Saint Louis. Her cross-species publications span topics of the biopsychosocial risk and protective factors of stress-related psychopathology. She is a great proponent of increasing diversity in psychological and neuroscience research and prides herself in her outreach activities to increase knowledge about the mind as well as the brain and its function. She is a contributing member of the Psychological Science Accelerator, and a member of Society for Research on Adolescents, Society for the Improvement of Psychological Science, Society for Neuroscience, and a faculty advisor of the first Mexican Psi Chi Chapter (University of Sonora).

Víctor Corral-Verdugo, PhD is a professor of psychology at the University of Sonora and expert in psychology of sustainability. Author of 10 books, editor of nine books, and author of more than 150 papers published in the USA, Europe, and Latin America, as well as book chapters (including chapters for the *Handbook of Environmental and Conservation Psychology*, 2012, Oxford, and *Research Methods for Environmental Psychology*, 2016, Wiley). His publications are cited by authors from more than 80 countries. He is also international associate editor of the journals *Environment and Behavior* (USA), *PsyEcology* (Spain), and *Frontiers in Psychology* (Switzerland). He has been awarded by APA for Best Paper on Environment and Population 1995, as a doctoral student, and received The Mexican Psychology Award, and the Interamerican Environmental Psychology Award in 2011.

Richard G. Cowden, PhD is a social-personality psychologist and the psychology research associate for the Human Flourishing Program at Harvard's Institute for Quantitative Social Science. He completed his doctoral training at the University of KwaZulu-Natal, South Africa. He is broadly interested in intersections between cultural-contextual dynamics and psychosocial processes that shape adaptive functioning, personal growth, and wellbeing. He has written more than 100 scholarly articles and book chapters that address various aspects of human flourishing in a wide range of populations, and recently co-authored a book titled *Place and Post-Pandemic Flourishing: Disruption, Adjustment, and Healthy Behaviors* (Springer, 2021).

Nerys Edmonds is an independent scholar and a principal public health practitioner in the Wales Health Impact Assessment Support Unit in Public Health Wales. Edmonds initially trained as a Mental Health Nurse and has had a passion for improving mental health and wellbeing and preventing mental health problems throughout her career. Since 2000, she has worked in public health in England and Wales in the NHS, local government, and third sector with a focus on promoting mental health and wellbeing and tackling health inequalities. This has included work on suicide prevention, promoting mental wellbeing across the life course, tackling stigma, early intervention, and health impact assessment (HIA). Edmonds is also an honorary lecturer in the School of Medicine in Cardiff University. Edmonds is an author of a number of publications including *Mental Well-being Impact Assessment: A Toolkit for Well-being* (2011), and the *Public Health Implications of Brexit in Wales: A Health Impact Assessment Approach.*

Darren J. Edwards, PhD is a BPS chartered health psychologist and a senior lecturer in the School of Health and Social Care at Swansea University. His academic interests encompass wellbeing science, categorisation models of cognition, data visualisation, psychological therapies such as ACT, the impact of these approaches within society, and how to utilise AI to identify predictors for ill health facilitating the development of early interventions. He has developed several policy briefs for the Welsh Government and in conjunction with the NHS in the area of health and AI. He has also recently published an edited textbook with SAGE called *The Textbook of Health and Social Care.*

Alecia Eubanks, PhD is an associate professor of international psychology at The Chicago School of Professional Psychology, USA. Eubanks began her early career as a Minority International Research Trainee in the U.S. National Institutes of Health. In this role, she lived in Cape Town, South Africa, researching and training teachers on best practices to educate middle school children on HIV prevention. In addition to living in South Africa, she has volunteered to teach English in Thailand and Costa Rica. Eubanks's professional path has embraced applied research, international services, educational programme administration, and student professionalisation. She believes deeply in creating one-on-one relationships with students, teaching transferable skills, and utilising creative methods to exchange knowledge.

Jen L. Fisher, PhD is a full-time lecturer in the Department of Psychology at the American University of Beirut (Lebanon) and Lead Monitoring & Evaluation Consultant and Disability Inclusion Expert at Cohort 22, LLC.

Zoe Fisher, PhD, DClinPsy Card is a consultant clinical psychologist and lead for Community Neuro-rehabilitation Services in Swansea Bay Health Board. She is also an associate professor at the Health and Wellbeing Academy at Swansea University. Her work focuses on developing theoretical frameworks and interventions that improve health and wellbeing and bridge the gap between research and clinical practice. Her work has been published and featured on the national media as an example of innovation in practice, and Fisher has received many awards in recognition of her work and has most recently received an Advancing Health Care Award for 'Outstanding Contribution to Research Delivery'. Prior to her doctorate training as a clinical psychologist at University of Wales, Cardiff, Fisher completed her PhD in neuropsychology at the University of Wales, Swansea. She has recently completed a Postgraduate Diploma in Clinical Neuropsychology at Bristol University.

Martha Frias-Armenta, PhD is a Professor at the Department of Law, University of Sonora, Mexico. She completed her law degree in the University of Sonora and her PhD in psychology and law in the University of Arizona and she is member of Mexico's National Researchers System. Her research interest focuses on antisocial and criminal behaviour, family violence, juvenile delinquency, restorative justice, procedural and distributive justice, and ecological crime. Moreover, she is an evaluator of research projects of science and technology for Mexican and Spain research councils and editor of Mexican and international scientific journals. She has been honoured as a distinguished researcher of the Social Sciences Division of her university. She has published seven books and several papers in national and international journals.

Katie Gibbs is a postgraduate researcher with an extensive background in wellbeing science and a history of working in applied healthcare. Her research focuses on using qualitative methodologies to understand how individuals with stroke and brain injury experience various innovative group-based wellbeing interventions (including surf therapy, sustainable building projects, and adapted acceptance and commitment therapy). Using evidence-based models as a lens through which participant insights may be understood and contextualised, her findings offer insights that extend upon the holistic model of neurorehabilitation, highlighting how community partnerships may be leveraged to create opportunities for wellbeing in those living with acquired brain injury.

Andrew H. Kemp, PhD, DH-SC Melb is a professor and personal chair in the School of Psychology at Swansea University in the United Kingdom. He has previously held posts at the University of Sao Paulo in Brazil (2013-2015) and the University of Sydney Australia (2004-2012). His academic background spans affective neuroscience through to epidemiology, and he currently teaches and conducts research in existential positive psychology, wellbeing science, and climate psychology. The significance of his work has been recognised in awards and international recognition including fellowship from British Psychological Society (2019) and the Association for Psychological Science in the United States (2017), as well as a Doctor of Science degree from the University of Melbourne (2019) in recognition of academic leadership, creative scholarship, and innovative contributions to the field.

Louise Lambert, PhD is the editor of the *Middle East Journal of Positive Psychology* and head of Happiness Policy and Programming with www.happinessmatters.org. She is a registered psychologist and Canadian national based in Dubai, where she has lived for more than 10 years. Her work and interests lie in positive education programming, the use of positive psychology interventions in the workplace, employee and managerial training, as well as content development for a range of wellbeing platforms across the Gulf region. She also teaches at university and conducts research on intervention programmes and other cultural and wellbeing outcomes.

Marc Yancy Lucas, PhD is an adjunct professor in the Psychology Department at the University of Sonora. He completed an undergraduate degree in anthropology at the University of Indiana and has a master's degree in Latin American studies and urban planning from the University of Arizona. He obtained his doctoral degree with honours in psychology at the University of Sonora, where his thesis focused on positive environments predicting prosocial driving behaviours. His research interests include environmental psychology, urbanism, and sustainability and their associations with wellbeing. He is a contributing member of the Psychological Science Accelerator and a former president and founding member of the Psi Chi chapter of the University of Sonora chapter.

Claude-Hélène Mayer, PhD is a professor in industrial and organisational psychology at the Department of Industrial Psychology and People Management at the University of Johannesburg, an adjunct professor at the European University Viadrina in Frankfurt (Oder), Germany. She holds doctoral degrees in Psychology, Management, and Cultural Anthropology and her *Venia Legendi* is in psychology with a focus on work, organisational, and cultural psychology. Her research areas are transcultural mental health and wellbeing, positive psychology and salutogenesis, transcultural conflict management and mediation, women in leadership in culturally diverse work contexts, and psychobiography. Her teaching areas include cross-cultural psychology, organisational theory, systems and design thinking, coaching, positive psychology, organisational behaviour, and transcultural conflict management and mediation.

David Neilson, PhD is a senior lecturer in the Department of Sociology and Social Policy at the University of Waikato, Hamilton, New Zealand. He completed his PhD in the field of social theory at the University of East Anglia. His current academic fields of interest include class theory, mid-range political economy, development studies, political philosophy, and social psychology. He recently published a non-sectarian and holistic contemporary reading, of Marx's works applied specifically to present challenges, titled *The Making of Democratic Socialism in the Twenty-First Century*. He is currently working on his new book for Routledge titled *Beyond Imperialism, Neoliberal Globalisation and Uneven Development: Towards a Democratic Socialist Alternative*. His well-received article 'Class, Precarity, and Anxiety under Neoliberal Global Capitalism: From Denial to Resistance' published in *Theory and Psychology* is particularly relevant to the chapter he has contributed to the present volume.

Dominic Page, PhD is the head of the Business School at the University of Gloucestershire, UK. A sociologist by training, Page's research expertise includes equality and diversity, welfare and welfare reform, mental health, and employment and workplace health. He has teaching interests in leadership, human resource management, management studies, and political economy. He has long-established education policy interests in internationalisation, challenge-based curriculum, the role of technology in enhancing teaching and learning, research-led teaching, and learning and industry partnership-based curriculum.

Alison Pritchard is a retired consultant in public health and former fellow of the UK Faculty of Public Health. During her career in public health, working for the NHS and local authorities in Derbyshire, she was actively involved in developing and leading public health interventions to promote the wellbeing benefits of nature at a population level, and served as a board member of the Derbyshire Peak District Local Nature Partnership. She is currently studying for her PhD in nature connectedness and eudaimonic wellbeing at Derby University.

Miles Richardson, PhD is a professor of human factors and nature connectedness and founded the Nature Connectedness Research Group at the University of Derby. The focus of this research is on understanding and improving connection with nature, given the wellbeing and environmental benefits. The group works closely with Natural England, and its work has been adopted by many organisations, including the National Trust, the Royal Society for the Protection of Birds (RSPB), and 2021 Mental Health Awareness Week. The group won the institutional award for research impact at the 2021 Green Gown Awards. Richardson is also a lead author on 'Intergovernmental Science-Policy Platform on Biodiversity and Ecosystem Services' (IPBES) global transformative change assessment.

Rachael J. Waldrop, MS received her graduate degree in Experimental Psychology from Western Washington University and her undergraduate degree in psychological sciences from Georgia College and State University. Her research focuses on the overlap of positive psychology and social groups, uncovering the ways in which outgroup members work together to overcome interpersonal and intergroup differences through value identification, education, and allyship.

Meg A. Warren, PhD is an associate professor of management at Western Washington University, USA. She is a researcher, keynote speaker, author, and psychologist with expertise in allyship, equity, inclusiveness, and cultural factors affecting wellbeing. Warren has published over 50 journal articles and book chapters. Her research has been covered by over 200 popular media outlets globally. She is the founding president of the Work & Organizations Division of the International Positive Psychology Association, co-founder of the Western Positive Psychology Association, and co-editor of the *International Journal of Wellbeing*. To access publications, media, and resources on her work, please visit www.megwarren.com.

Don J. Webber, PhD is a professor of managerial economics at The University of Sheffield, UK. He carries out policy-relevant, social science research with a multidisciplinary and applied focus. Webber has won grants from UK Research Councils, UK government departments, government departments overseas, and charities. His work has been discussed at the United Nations' International Labour Organization (ILO), the Central Bank of Nigeria, and elsewhere. His teaching history encompasses a variety of topics at all university levels and his teaching quality has been recognised by students and the wider academic community including being awarded a Special Commendation for 'Outstanding Teaching' in the UK's LTSN national competition. Webber's first book is *How to Enhance Your Research: 100 Practical Tips for Academics* published by Edward Elgar in 2021 (ISBN: 9781788978088).

Paul T. P. Wong, PhD, CPsych is professor emeritus of Trent University. He is a fellow of APA and the Canadian Psychological Association (CPA) and president of the International Network on Personal Meaning (www.meaning.ca) and the Meaning-Centered Counselling Institute Inc. Editor of the *International Journal of Existential Positive Psychology*, he has also edited two influential volumes of *The Human Quest for Meaning*. A prolific writer, he is one of the most cited existential and positive psychologists. The originator of meaning therapy and International Meaning Conferences, he has been invited to give keynotes and meaning therapy workshops worldwide. He is the recent recipient of the Carl Rogers Award from the Society for Humanistic Psychology.

List of Figures

CHAPTER 1

An Introduction to the Complex Construct of Wellbeing, Societal Challenges and Potential Solutions

Andrew H. Kemp and Darren J. Edwards

Abstract This chapter provides a brief introduction to the complex construct of wellbeing, some of the major societal challenges impacting on it and potential solutions. Wellbeing is a controversial topic and its definition is typically dependent on disciplinary perspectives. Here we take a broader view, encompassing individual, collective and planetary wellbeing that focuses on individuals, groups and the systems in which they are embedded. Some major barriers to achieving a sense of wellbeing include problems within society such as chronic disease, poverty and loneliness, intensified

A. H. Kemp (✉)
School of Psychology, Faculty of Medicine, Health and Life Science, Swansea University, Swansea, Wales, UK
e-mail: a.h.kemp@swansea.ac.uk

D. J. Edwards
Department of Public Health, School of Health and Social Care, Faculty of Medicine, Health and Life Science, Swansea University, Swansea, Wales, UK
e-mail: D.J.Edwards@Swansea.ac.uk

A. H. Kemp, D. J. Edwards (eds.), *Broadening the Scope of Wellbeing Science*, https://doi.org/10.1007/978-3-031-18329-4_1

by the climate crisis and reinforced by the neoliberal model of development, all of which impact on the most vulnerable in society. Progress on sustainable development goals has been disappointing, and there is an urgent need for transformational change at multiple levels of scale that focuses on qualities and skills for inner development that lay stronger foundations for improving our capacity to manage increasingly complex societal challenges. Subsequent chapters further explore multidisciplinary and inter-disciplinary issues relating to understanding and promotion of wellbeing despite great hardship, suffering and inequity.

Keywords Wellbeing • Positive psychology • Contextual behavioural science • Inner development goals

Dictionary definitions of 'wellbeing' refer to the state of being comfortable, happy and healthy, suggesting that wellbeing encompasses positive emotions, positive feelings and positive health. Similarly, the constitution of the World Health Organization (WHO) defines 'health' as an overarching construct that includes complete mental, physical and social wellbeing. These definitions are inspiring, but they also dismiss inevitable suffering experienced by everyone over a lifetime—more or less—impacting on individual capacities to experience wellbeing. Psychological scientists have approached the construct of wellbeing from a different perspective, emphasising hedonia (e.g. positive emotions), eudaimonia (e.g. meaning) and concepts of flourishing relating to a balance of both (Diener, 1984, 2000; Ryff, 1989; Ryff & Keyes, 1995; Seligman, 2012, 2017). However, psychological models typically ignore wider contextual factors and societal challenges including structural inequities, environmental degradation and the climate crisis, and have been criticised accordingly (Carlisle et al., 2009; Davies, 2015; Frawley, 2015). Additional factors including access to education, social equality, social trust and cohesion, civic engagement, and the quality of physical environments amongst others critically impact on wellbeing (Delhey & Dragolov, 2016; Putnam, 2000; Tran et al., 2021; Ulrich et al., 1991; Ward & Meyer, 2009; Wilkinson & Pickett, 2010).

Our understanding of what it means to experience wellbeing, however, is evolving. Complementary ideas emerging in positive psychology and contextual behavioural science illustrate how the discipline of psychological science is maturing (Atkins et al., 2019; Ciarrochi et al., 2022; Diener et al., 2020; Kern et al., 2020; Lomas et al., 2020; Wong, 2019), focused on the integration of previously isolated fields and a move beyond the individual toward impacts at higher levels of scale. The editors' own work in this area links awareness of one's own bodily feelings and functioning of the vagus nerve to psychological flexibility, emotional regulation and wellbeing (e.g. Kemp, Koenig, et al., 2017; Pinna & Edwards, 2020), linking vagal functioning to wellbeing in a multileveled theoretical framework that places the individual within the context of their social and natural ecologies (Kemp & Fisher, 2022; Mead et al., 2021). This work seeks to better understand the relationship between mental and physical functioning, and that of the groups and systems within which people are embedded. Edwards has described how relational frame theory—a theory of language based on contextual relationships on which acceptance and commitment therapy has been developed—may provide foundations for more effective public health communication (Edwards, 2021), while Kemp has sought to better understand the links between often contradictory constructs relating to the construct of wellbeing (e.g. happiness and suffering) (Gibbs et al., 2022; Wilkie et al., 2021) and the relevance of wellbeing to tackling major societal challenges such as the increasing burden of chronic disease (Kemp et al., 2022; Kemp & Fisher, 2022). This work is aligned with efforts that seek to promote health and wellbeing at scale (e.g. application of a 'prosocial' framework to facilitate collaboration, cooperation and care within and between groups) (Atkins et al., 2019), alongside a complimentary emerging transdisciplinary effort to improve 'planetary wellbeing', which has been defined as the "the highest attainable standard of wellbeing for human and non-human beings and their social and natural systems" (Antó et al., 2021). Kemp and colleagues' GENIAL model of wellbeing imposes an interpretative framework on often contradictory developments, emphasising 'connection' to ourselves, community and the natural environment as well as the need for positive change at multiple levels of scale to address sociostructural barriers that constrain wellbeing (Fisher et al., 2020; Kemp, Arias, et al., 2017; Mead et al., 2021).

This background and context provided the foundations on which we invited colleagues working in the heterogeneous field of wellbeing science to contribute chapters to this book, providing a range of overlapping and often critical perspectives that we hope will have a lasting impact on the field for years to come. We now briefly describe some of the societal challenges to and potential solutions emerging from the field in recent years, providing useful context for reading of subsequent chapters.

Societal Challenges to Wellbeing

Challenges include increasing loneliness (Buecker et al., 2021), and psychiatric morbidity (GBD 2019 Mental Disorders Collaborators, 2022) with impacts equivalent to the effects of heavy smoking (Chesney et al., 2014; Holt-Lunstad et al., 2010); rising social and health-related inequalities, fuelling discontent, political dysfunction and authoritarian populism (Jay et al., 2019; Marmot, 2015, 2020; Oxendine, 2019; United Nations World Social Report, 2020), the spread of misinformation and state-sponsored disinformation, contributing to societal tensions and violent conflict (Lewandowsky et al., 2013; Silverman et al., 2021), and the unfolding climate catastrophe (Ripple et al., 2017, 2019, 2021), which is arguably intensifying and multiplying all other challenges (Li et al., 2021; Pörtner et al., 2022), a catastrophe that is reinforced by dominant Western neoliberal ideology with profound psychological roots (Weintrobe, 2021). Societal challenges are interconnected, and it is helpful to frame these within the context of the documented 'systemic risk' of climate change (Li et al., 2021), arising through the complex interaction of climate and complex human systems such as global food, energy and financial markets, with major implications for health and wellbeing (Morgan et al., 2022). Research has highlighted (Semieniuk et al., 2022) the challenge associated with planning for a deep and fast structural transition to a global low-carbon economy, with major implications for private investors including substantial exposure through pension funds and financial markets. This financial risk of stranded fossil fuel assets (estimated at US$681 billion) far exceeds that associated with mispriced subprime housing assets (US$250–500 billion), responsible for the 2007–2008 financial crisis. The human cost of this transition will be substantial, and the mental health impacts predicted in Intergovernmental Panel on Climate Change (IPCC) reports are already beginning to materialise through reports of increasing psychological distress in children and young people (Hickman et al., 2021), the rise of the 'BirthStrike' movement (McMullen & Dow, 2022) and increased risk of climate-related suicide (Dumont et al., 2020).

Pressing global challenges led the United Nations to propose the sustainable development goals in 2015 (UNSDGs, https://sdgs.un.org/), concentrating on human, social and environmental development objectives with aspirations to achieve those goals by 2030. These included 17 goals such as good health and wellbeing, no poverty, reduced inequalities, responsible consumption and climate action amongst others, focused on 169 targets and 232 specific indicators. Unfortunately progress on these goals has been disappointing (Moyer & Hedden, 2020), arguably arising from the lack of inner capacity to manage increasingly complex challenges (Wamsler & Brink, 2018) [see also: https://www.innerdevelopmentgoals.org/], reinforcing a key role for psychological science in a coordinated inter-disciplinary response to promote transformational change.

Emerging Solutions to Societal Challenges

Emerging solutions highlight a need for promoting 'inner development goals' (or IDGs) with the aim of improving our capacity to reach the UNSDGs [https://www.innerdevelopmentgoals.org/] underpinned by the work of Christine Wamsler and others (Wamsler & Brink, 2018; Woiwode et al., 2021). While this initiative is still evolving, an initial framework has identified 23 skills and qualities across five categories encompassing 'Being', 'Thinking', 'Relating', 'Collaborating' and 'Acting' [see: https://www.innerdevelopmentgoals.org/]. This work complements other developments and ideas in psychological science focused on wellbeing, which has been characterised as shift in focus to multiple levels of scale that encompass individuals, groups and systems in which people are embedded (Atkins et al., 2019; Kern et al., 2020; Lomas et al., 2020; Mead et al., 2019, 2021). Against the background of the unfolding climate catastrophe, researchers have begun to explore the significance of climate-related emotions and implications for health and wellbeing, resilience, pro-environmental behaviours, climate action, morality and ethics (Morgan et al., 2022; Pihkala, 2022). Surprisingly, while climate-related emotions include numerous variants of negative experience such as climate-related grief, anxiety, anger and rage, the crisis can also trigger a range of positive emotions, including love, care, hope and empowerment (Pihkala, 2022), promoted through, for example, Joanna Macy's "work that reconnects" (Macy & Brown, 2014; Macy & Johnstone, 2012). These developments and emerging solutions motivated us to solicit book chapters from colleagues whose work has influenced our own thinking on the topic of wellbeing, and to reflect on how the field of wellbeing science might move forward. The following chapters provide some guidance for the path forward including a reimagining of what it means to be human and the need to embrace and transform the inevitable experience of suffering (Chap. 2) with a broader focus on culture, social interactions and climate action (Chap. 3), through application of psychological science at increasingly higher levels of scale (Chaps. 3, 4 and 5), a focus on the natural environment (Chaps. 6 and 7), and contributions from beyond psychological science (Chaps. 8, 9 and 10) highlighting an urgent need for enhanced cross-disciplinary engagement. Together this work contributes to a broader understanding of the complex construct of wellbeing, laying the foundations for future solutions, focused on building the needed skills and qualities to mitigate and adapt to the systemic risk posed by climate change and the many interconnected challenges that must now be addressed by current and future generations.

References

Antó, J. M., Martí, J. L., Casals, J., Bou-Habib, P., Casal, P., Fleurbaey, M., Frumkin, H., Jiménez-Morales, M., Jordana, J., Lancelotti, C., Llavador, H., Mélon, L., Solé, R., Subirada, F., & Williams, A. (2021). The planetary wellbeing initiative: Pursuing the sustainable development goals in higher education. *Sustainability, 13*(6), 3372. https://doi.org/10.3390/su13063372

Atkins, P. W. B., Wilson, D. S., & Hayes, S. C. (2019). *Prosocial: Using evolutionary science to build productive, equitable, and collaborative groups.* New Harbinger Publications.

Buecker, S., Mund, M., Chwastek, S., Sostmann, M., & Luhmann, M. (2021). Is loneliness in emerging adults increasing over time? A preregistered cross-temporal meta-analysis and systematic review. *Psychological Bulletin, 147*(8), 787–805. https://doi.org/10.1037/bul0000332

Carlisle, S., Henderson, G., & Hanlon, P. W. (2009). 'Wellbeing': A collateral casualty of modernity? *Social Science & Medicine, 69*(10), 1556–1560. https://doi.org/10.1016/j.socscimed.2009.08.029

Chesney, E., Goodwin, G., & Fazel, S. (2014). Risks of all-cause and suicide mortality in mental disorders: A meta-review. *World Psychiatry, 13*(153), 160.

Ciarrochi, J., Hayes, S. C., Oades, L. G., & Hofmann, S. G. (2022). Toward a unified framework for positive psychology interventions: Evidence-based processes of change in coaching, prevention, and training. *Frontiers in Psychology, 12*, 809362. https://doi.org/10.3389/fpsyg.2021.809362

GBD 2019 Mental Disorders Collaborators. (2022). Global, regional, and national burden of 12 mental disorders in 204 countries and territories, 1990–2019: A systematic analysis for the global burden of disease study 2019. *The Lancet Psychiatry, 9*(2), 137–150. https://doi.org/10.1016/s2215-0366(21)00395-3

Davies, W. (2015). *The happiness industry.* Verso Books.

Delhey, J., & Dragolov, G. (2016). Happier together. Social cohesion and subjective well-being in Europe. *International Journal of Psychology, 51*(3), 163–176. https://doi.org/10.1002/ijop.12149

Diener, E. (1984). Subjective well-being. *Psychological Bulletin, 95*(3), 542. https://doi.org/10.1037/0033-2909.95.3.542

Diener, E. (2000). Subjective well-being: The science of happiness and a proposal for a national index. *American Psychologist, 55*(1), 34–43. https://doi.org/10.1037//0003-066x.55.1.34

Diener, E., Miyata, H., Kawakami, T., Chen, D., Kitagawa, T., Lai, A., Ishikawa, Y., Harter, J., Joshanloo, M., van de Weijer, M., Lambert, L., Passmore, H.-A., & Lomas, T. (2020). Towards a greater global understanding of wellbeing: A proposal for a more inclusive measure. *International Journal of Wellbeing, 10*(2), 1–18. https://doi.org/10.5502/ijw.v10i2.1037

Dumont, C., Haase, E., Dolber, T., Lewis, J., & Coverdale, J. (2020). Climate change and risk of completed suicide. *Journal of Nervous & Mental Disease, 208*(7), 559–565. https://doi.org/10.1097/nmd.0000000000001162

Edwards, D. J. (2021). Ensuring effective public health communication: Insights and modeling efforts from theories of behavioral economics, heuristics, and behavioral analysis for decision making under risk. *Frontiers in Psychology, 12*, 715159. https://doi.org/10.3389/fpsyg.2021.715159

Fisher, Z., Galloghly, E., Boglo, E., Gracey, F., & Kemp, A. (2020). Emotion, wellbeing and the neurological disorders. In *Reference Module in Neuroscience and Biobehavioral Psychology*. Elsevier. https://doi.org/10.1016/b978-0-12-819641-0.00013-x.

Frawley, A. (2015). Happiness research: A review of critiques. *Sociology Compass, 9*(1), 62–77. https://doi.org/10.1111/soc4.12236

Gibbs, K., Wilkie, L., Jarman, J., Barker-Smith, A., Kemp, A. H., & Fisher, Z. (2022). Riding the wave into wellbeing: A qualitative evaluation of surf therapy for individuals living with acquired brain injury. *PLoS One, 17*(4), e0266388. https://doi.org/10.1371/journal.pone.0266388

Hickman, C., Marks, E., Pihkala, P., Clayton, S., Lewandowski, R. E., Mayall, E. E., Wray, B., Mellor, C., & van Susteren, L. (2021). Climate anxiety in children and young people and their beliefs about government responses to climate change: A global survey. *The Lancet Planetary Health, 5*(12), e863–e873. https://doi.org/10.1016/s2542-5196(21)00278-3

Holt-Lunstad, J., Smith, T. B., & Layton, J. B. (2010). Social relationships and mortality risk: A meta-analytic review. *PLoS Medicine, 7*(7), e1000316. https://doi.org/10.1371/journal.pmed.1000316

Jay, S., Batruch, A., Jetten, J., McGarty, C., & Muldoon, O. T. (2019). Economic inequality and the rise of far-right populism: A social psychological analysis. *Journal of Community & Applied Social Psychology, 29*(5), 418–428. https://doi.org/10.1002/casp.2409

Kemp, A. H., Arias, J. A., & Fisher, Z. (2017). Social ties, health and wellbeing: A literature review and model. In *Neuroscience and Social Science, the Missing Link* (Vol. 59, pp. 397–427). Springer International Publishing. https://doi.org/10.1007/978-3-319-68421-5_17.

Kemp, A. H., & Fisher, Z. (2022). Wellbeing, whole health and societal transformation: Theoretical insights and practical applications. *Global Advances in Health and Medicine, 11*, 21649561211073076. https://doi.org/10.1177/21649561211073077

Kemp, A. H., Koenig, J., & Thayer, J. (2017). From psychological moments to mortality: A multidisciplinary synthesis on heart rate variability spanning the continuum of time. *Neuroscience and Biobehavioral Reviews, 83*, 547–567. https://doi.org/10.1016/j.neubiorev.2017.09.006

Kemp, A. H., Tree, J., Gracey, F., & Fisher, Z. (2022). Editorial: Improving wellbeing in patients with chronic conditions: Theory, evidence, and opportunities. *Frontiers in Psychology, 13*, 868810. https://doi.org/10.3389/fpsyg.2022.868810

Kern, M. L., Williams, P., Spong, C., Colla, R., Sharma, K., Downie, A., Taylor, J. A., Sharp, S., Siokou, C., & Oades, L. G. (2020). Systems informed positive psychology. *The Journal of Positive Psychology, 15*(6), 705–715. https://doi.org/10.1080/17439760.2019.1639799

Lewandowsky, S., Stritzke, W. G. K., Freund, A. M., Oberauer, K., & Krueger, J. I. (2013). Misinformation, disinformation, and violent conflict. *American Psychologist, 68*(7), 487–501. https://doi.org/10.1037/a0034515

Li, H.-M., Wang, X.-C., Zhao, X.-F., & Qi, Y. (2021). Understanding systemic risk induced by climate change. *Advances in Climate Change Research, 12*(3), 384–394. https://doi.org/10.1016/j.accre.2021.05.006

Lomas, T., Waters, L., Williams, P., Oades, L. G., & Kern, M. L. (2020). Third wave positive psychology: Broadening towards complexity. *The Journal of Positive Psychology, 16*, 1–15. https://doi.org/10.1080/17439760.2020.1805501

Macy, J., & Brown, M. Y. (2014). *Coming back to life: The updated guide to the work that reconnects*. New Society Publishers.

Macy, J., & Johnstone, C. (2012). *Active hope: How to face the mess we're in without going crazy*. New World Library.

Marmot, M. (2015). The health gap: The challenge of an unequal world. *Lancet, 386*(10011), 2442–2444. https://doi.org/10.1016/s0140-6736(15)00150-6

Marmot, M. (2020). Society and the slow burn of inequality. *The Lancet, 395*(10234), 1413–1414. https://doi.org/10.1016/s0140-6736(20)30940-5

McMullen, H., & Dow, K. (2022). Ringing the existential alarm: Exploring BirthStrike for climate. *Medical Anthropology, 41*, 659–673. https://doi.org/10.1080/01459740.2022.2083510

Mead, J., Fisher, Z., & Kemp, A. H. (2021). Moving beyond disciplinary silos towards a transdisciplinary model of wellbeing: An invited review. *Frontiers in Psychology, 12*, 642093. https://doi.org/10.3389/fpsyg.2021.642093

Mead, J., Fisher, Z., Wilkie, L., Gibbs, K., Pridmore, J., Tree, J., & Kemp, A. (2019). Toward a more ethical science of wellbeing that considers current and future generations. *Authorea*. https://doi.org/10.22541/au.156649190.08734276

Morgan, G., Barnwell, G., Johnstone, L., Shukla, K., & Mitchell, A. (2022). The power threat meaning framework and the climate and ecological crises. *Psychology in Society, 63*, 83–109.

Moyer, J. D., & Hedden, S. (2020). Are we on the right path to achieve the sustainable development goals? *World Development, 127*, 104749. https://doi.org/10.1016/j.worlddev.2019.104749

Oxendine, A. R. (2019). The political psychology of inequality and why it matters for populism. *International Perspectives in Psychology: Research, Practice, Consultation, 8*(4), 179–195. https://doi.org/10.1037/ipp0000118

Pihkala, P. (2022). Toward a taxonomy of climate emotions. *Frontiers in Climate, 3*, 738154. https://doi.org/10.3389/fclim.2021.738154

Pinna, T., & Edwards, D. J. (2020). A systematic review of associations between interoception, vagal tone, and emotional regulation: Potential applications for mental health, wellbeing, psychological flexibility, and chronic conditions. *Frontiers in Psychology, 11*, 1792. https://doi.org/10.3389/fpsyg.2020.01792

Pörtner, H.-O., Roberts, D. C., Tignor, M., Poloczanska, E. S., Mintenbeck, K., Alegría, A., Craig, M., Langsdorf, S., Löschke, S., Möller, V., Okem, A., & Rama, B. (Eds.). (2022). *IPCC, 2022: Climate change 2022: Impacts, adaptation, and vulnerability. Contribution of working group II to the sixth assessment report of the intergovernmental panel on climate change.* Cambridge University Press.

Putnam, R. D. (2000). *Bowling alone*. Simon & Schuster. http://bowlingalone.com/.

Ripple, W. J., Wolf, C., Newsome, T. M., Barnard, P., & Moomaw, W. R. (2019). World scientists' warning of a climate emergency. *Bioscience, 70*, 8–12. https://doi.org/10.1093/biosci/biz088

Ripple, W. J., Wolf, C., Newsome, T. M., Galetti, M., Alamgir, M., Crist, E., Mahmoud, M. I., & Laurance, W. F. (2017). World scientists' warning to humanity: A second notice. *Bioscience, 67*(12), 1026–1028. https://doi.org/10.1093/biosci/bix125

Ripple, W. J., Wolf, C., Newsome, T. M., Gregg, J. W., Lenton, T. M., Palomo, I., Eikelboom, J. A. J., Law, B. E., Huq, S., Duffy, P. B., & Rockström, J. (2021). World scientists' warning of a climate emergency 2021. *Bioscience, 71*, 894. https://doi.org/10.1093/biosci/biab079

Ryff, C. D. (1989). Happiness is everything, or is it? Explorations on the meaning of psychological well-being. *Journal of Personality and Social Psychology, 57*(6), 1069–1081.

Ryff, C. D., & Keyes, C. L. M. (1995). The structure of psychological well-being revisited. *Journal of Personality and Social Psychology, 69*(4), 719–727.
Seligman, M. (2012). *Flourish: A visionary new understanding of happiness and well-being.* Free Press.
Seligman, M. (2017). PERMA and the building blocks of well-being. *The Journal of Positive Psychology, 13*(4), 1–3. https://doi.org/10.1080/17439760.2018.1437466
Semieniuk, G., Holden, P. B., Mercure, J.-F., Salas, P., Pollitt, H., Jobson, K., Vercoulen, P., Chewpreecha, U., Edwards, N. R., & Viñuales, J. E. (2022). Stranded fossil-fuel assets translate to major losses for investors in advanced economies. *Nature Climate Change, 12*, 1–7. https://doi.org/10.1038/s41558-022-01356-y
Silverman, D., Kaltenthaler, K., & Dagher, M. (2021). Seeing is disbelieving: The depths and limits of factual misinformation in war. *International Studies Quarterly, 65*(3), 798–810. https://doi.org/10.1093/isq/sqab002
Tran, D. B., Pham, T. D. N., & Nguyen, T. T. (2021). The influence of education on women's well-being: Evidence from Australia. *PLoS One, 16*(3), e0247765. https://doi.org/10.1371/journal.pone.0247765
Ulrich, R. S., Simons, R. F., Losito, B. D., Fiorito, E., Miles, M. A., & Zelson, M. (1991). Stress recovery during exposure to natural and urban environments. *Journal of Environmental Psychology, 11*, 201–230.
United Nations World Social Report. (2020). *World social report 2020: Inequality in a rapidly changing world.* United Nations: Department of Economic and Social Affairs.
Wamsler, C., & Brink, E. (2018). Mindsets for sustainability: Exploring the link between mindfulness and sustainable climate adaptation. *Ecological Economics, 151*, 55–61. https://doi.org/10.1016/j.ecolecon.2018.04.029
Ward, P., & Meyer, S. (2009). Trust, social quality and wellbeing: A sociological exegesis. *Development and Society, 38*(2), 339–363. https://doi.org/10.2307/deveandsoci.38.2.339
Weintrobe, S. (2021). *Psychological roots of the climate crisis: Neoliberal exceptionalism and the culture of uncare.* Bloomsbury Publishing. https://books.google.co.uk/books/about/Psychological_Roots_of_the_Climate_Crisi.html?id=ez0ZEAAAQBAJ&source=kp_book_description&redir_esc=y.
Wilkie, L., Arroyo, P., Conibeer, H., Kemp, A. H., & Fisher, Z. (2021). The impact of psycho-social interventions on the wellbeing of individuals with acquired brain injury during the COVID-19 pandemic. *Frontiers in Psychology, 12*, 648286. https://doi.org/10.3389/fpsyg.2021.648286
Wilkinson, R. G., & Pickett, K. (2010). *The spirit level: Why greater equality makes societies stronger.* Bloomsbury Press.

Woiwode, C., Schäpke, N., Bina, O., Veciana, S., Kunze, I., Parodi, O., Schweizer-Ries, P., & Wamsler, C. (2021). Inner transformation to sustainability as a deep leverage point: Fostering new avenues for change through dialogue and reflection. *Sustainability Science, 16*(3), 841–858. https://doi.org/10.1007/s11625-020-00882-y

Wong, P. T. P. (2019). Second wave positive psychology's (PP 2.0) contribution to counselling psychology. *Counselling Psychology Quarterly, 32*, 1–10. https://doi.org/10.1080/09515070.2019.1671320

CHAPTER 2

Shifting the Paradigm of Positive Psychology: Toward an Existential Positive Psychology of Wellbeing

Paul T. P. Wong, *Richard G. Cowden*,
Claude-Hélène Mayer, *and Victoria L. Bowers*

Abstract Recent developments in positive psychology and wellbeing research highlight the tremendous capacity for individual wellbeing within the context of a cross-cultural, ecological, and complex model. In this chapter, we propose a paradigm shift toward existential positive

---

We have no known conflict of interest to disclose.

P. T. P. Wong (✉)
Department of Psychology, Trent University, Peterborough, ON, Canada

R. G. Cowden
Human Flourishing Program, Institute for Quantitative Social Science, Harvard University, Cambridge, MA, USA

A. H. Kemp, D. J. Edwards (eds.), *Broadening the Scope of Wellbeing Science*, https://doi.org/10.1007/978-3-031-18329-4_2

psychology, which integrates the dark (negative psychology) and the bright (positive psychology) sides of life to provide a more integrative and comprehensive model of wellbeing. We explore the dialectical interaction between these poles of human experience, with a particular emphasis on the importance of embracing and transforming suffering for cultivating sustainable wellbeing. We highlight some of the innovative ways in which existential positive psychology differs from traditional positive psychology and summarize self-transcendence as the thread that runs through the existential positive psychology paradigm. Our chapter concludes with a brief discussion of some potential implications of existential positive psychology for research and interventions.

Keywords Existential positive psychology • Positive psychology • Self-transcendence • Suffering • Wellbeing

Introduction

Times and conditions have changed significantly since positive psychology emerged as a mainstream subfield of psychology in the early 2000s. The coronavirus disease 2019 (COVID-19) pandemic has catastrophically awoken us to the fragility of modern society and the finitude of the human condition (Counted et al., 2021; Cowden, Rueger, et al., 2021), and its impacts on human life have been widespread and persistent (De Kock et al., 2022; Govender et al., 2020; Jacobi et al., 2022; Shiba, Cowden, Counted, et al., 2022; Wong, Arslan, et al., 2021). The recent Russia-Ukraine war threatens to become a global conflict, with wide-ranging implications around the world from financial stress caused by rises in the price of energy through to the loss of civilian loved ones who have been collateral damage in the war. In response to these troubling circumstances, people may ask, "How can we be happy in times like this?" Traditional positive psychology does not provide an adequate response to this

C.-H. Mayer
Department of Industrial Psychology and People Management, University of Johannesburg, Johannesburg, South Africa

V. L. Bowers
Graduate College, Saybrook University, San Francisco, CA, USA

question, because it is principally focused on "understanding how people rise from zero to positive eight" (Gable & Haidt, 2005, p. 103). In contrast, existential positive psychology is fundamentally concerned with the question of how to stay positive and live well even in the darkest of hours amid great tragedy (Wong, 2021; Wong, Mayer, et al., 2021). This is an important consideration, as psychologists have a professional and moral obligation to address people's suffering and help them find hope for a better future. In this chapter, we advance a paradigm shift from traditional positive psychology by presenting an existential positive psychology vision that learning how to transform suffering is the beginning of sustainable wellbeing.

The Universality of Suffering

It is not possible to live a life without experiencing some form of suffering. This is not only a truism that has been the subject of much discourse during human history (Cowden, Davis, et al., 2021; Ho et al., 2022), but one with empirical support (Fowers et al., 2017; Gruber et al., 2011; Wong, 2019; Wong, Mayer, et al., 2021). There is also a vast body of literature on the adverse effects of existential concerns, distress, and suffering (Wong & Yu, 2021; Yalom, 1980). The COVID-19 pandemic has accelerated the urgency with which we need to address universal existential suffering (Bland, 2020; Mayer, 2021) and recognize the perils of pursuing happiness while neglecting suffering (Wong, 2011; Zerwas & Ford, 2021).

In Buddhist philosophy, the first noble truth is that life is suffering because cravings for worldly happiness and ignorance of the impermanence of life are the major sources of suffering (Cowden, Counted, et al., 2021; Targ & Hurtak, 2006). Yet, there is now a critical mass of empirical evidence about the upside of negative emotions and the downside of positive emotions (Friedman & Robbins, 2012; Gruber et al., 2011; Kashdan & Biswas-Diener, 2014; Mayer & Vanderheiden, 2019, 2021; Wong, 2019). This suggests that wellbeing needs actions oriented toward enhancing happiness (e.g., the Yang, or white half of the Tai-chi symbol with a small black circle near the center) as well as actions that facilitate the transformation of suffering (e.g., the Yin, or black half of the Tai-chi symbol with a small white circle near the center). Sustainable wellbeing can only be achieved by learning how to make the best use of the dynamic and dialectic interplay between positive and negative life experiences. The ancient Yin-Yang dialectic or the contemporary dual-system model

(Wong, 2012) provides a blueprint for how to navigate between opposite forces, such as good and evil or self and other, which are prevalent in life (Deng et al., 2020; Lomas & Ivtzan, 2016; Wong, 2020; Wong & Bowers, 2019).

New Developments in Wellbeing Research Within Psychology

There is a movement toward a more comprehensive conceptualization, measurement, and promotion of human wellbeing (Kemp & Fisher, 2022; Lee et al., 2021; Shiba, Cowden, Gonzalez, et al., 2022; VanderWeele, 2017). Although wellbeing is complex and irreducible to a single dimension, existential wellbeing is among the most important dimensions of complete wellbeing, because our ability to flourish depends on how well we respond to the reality that life involves suffering. Until recently, this reality has largely been overlooked within the positive psychology subfield.

Positive psychology is undergoing a period of great change, and this change has been captured in the metaphor of waves (Lomas et al., 2021). On this view, "psychology as usual" (negative psychology) is characterized by a focus on dysfunction and pathology; this was the thesis. Its antithesis was the first wave of positive psychology, with a principal focus on the positive. The synthesis was the second wave of positive psychology, characterized by a more nuanced approach toward the positive and negative by emphasizing the importance of harmonizing opposing elements into an integrative whole. The third wave incorporated knowledge and methodologies from diverse fields, focusing on groups, organizations, cultures, and systems in which people and their wellbeing are embedded.

In favor of Occam's razor, or the law of parsimony, we conceptualize these changes in terms of a paradigm shift because it is more consistent with the enduring notion of scientific revolutions (Kuhn, 1962). Moreover, whereas the wave metaphor reflects gradual change, the paradigm shift concept signals that there are fundamental changes in assumptions, theory, research, and intervention. Using this framing, negative psychology might represent the thesis and positive psychology the antithesis. Existential positive psychology is characterized by the synthesis of the thesis and antithesis through a process of transcending polarity (see Fig. 2.1), while also capturing the contextual and cultural complexities outlined in the third wave put forward by Lomas et al. (2021).

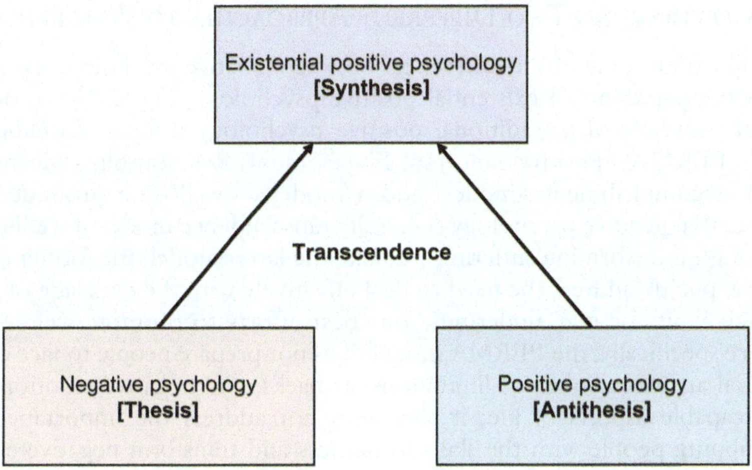

Fig. 2.1 A paradigm shift toward an existential positive psychology synthesizing negative and positive psychology

A paradigm shift toward an existential positive psychology is advantageous because it offers an integrative framework for understanding, measuring, and promoting wellbeing. For example, whereas traditional positive psychology tends to overlook the important role that suffering plays in human experience (see Coyne & Tennen, 2010; Held, 2004; Lazarus, 2003; Miller, 2008; Ryff, 2022; Wong, 2019, 2021), existential positive psychology addresses both the bright and dark sides of life in a more holistic way (Mayer, 2020; Mayer & Vanderheiden, 2021; Wong, 2017). The paradigm of existential positive psychology also represents a fundamental shift in both attitude and action; learning how to accept and embrace suffering can be one of the most promising paths to sustainable wellbeing (Kjell, 2011; Wong & Bowers, 2019). From this perspective, suffering may contribute to wellbeing by (a) prompting people to search for meaning as a more favorable option than responses that lead to maladaptive consequences (e.g., substance abuse, aggression, suicide), (b) teaching people how to manage or overcome painful emotions and experiences so that they have a greater likelihood of staying happy and healthy, and (c) triggering diagnostic, curative, redemptive, or growth-promoting responses that serve transformative functions (Wong, Mayer, et al., 2021).

Contrasting Two Different Approaches to Wellbeing

To illustrate some of the differences between positive psychology as usual and the paradigm of existential positive psychology, Fig. 2.2 provides a brief overview of a traditional positive psychology model of wellbeing (i.e., PERMA—Positive Emotion, Engagement, Relationships, Meaning, and Accomplishment—model) and a model of wellbeing grounded in existential positive psychology (i.e., self-transcendence model of wellbeing through transforming suffering). Unlike the latter model, the former does not explicitly address the need to deal effectively with the dark side of life, which limits or may undermine our best efforts to promote wellbeing. More specifically, the PERMA model does not prepare people to face existential anxieties, inherent limitations, setbacks, and negative emotions as inescapable aspects of life; it also does not address the importance of equipping people with the skills to harness and transform negative emotions and experiences into positive energies that can enhance wellbeing. For example, the pursuit and expectation of positive emotions often lead to disappointment and unhappiness because it is not possible to avoid negative emotions, which are not only natural responses to some situations but are considered necessary for our wellbeing (Hershfield et al., 2013).

Existential positive psychology is therefore a positive psychology that recognizes, embraces, and integrates the dark side of human existence, providing an existential dimension to personhood. The four tenets of this existential positive psychology paradigm can be summarized as follows: (a) polarity, such that all things in nature exist as two opposite but complementary dimensions; (b) self-transcendence, referring to a process of rising above limitations and conflicts by navigating a balanced and dynamic dialectic between opposites; (c) balance and harmony, capturing the idea that inner peace and wellbeing result from balancing the different and

Constituents	Seligman's (2011) model of wellbeing	Wong's (2021) model of wellbeing
P	Positive emotions	Positivity through embracing the dark side of life
E	Engagement	Engagement through painful discipline
R	Positive relationships	Strong relationships through forgiving and enduring the foibles of others
M	Meaning	Meaning through transcending egotism and external obstacles
A	Accomplishments	Achievement through sacrifices and overcoming failures

Fig. 2.2 Comparison between Seligman's (2011) model of wellbeing and Wong's (2021) self-transcendence model of wellbeing through transforming suffering

often competing dimensions within the individual (e.g., physical, mental, spiritual); and (d) true positivity, which involves seeing or being the light in times of darkness or suffering. From this perspective, positivity is not defined in terms of a positive to negative ratio but is based on one's ability to transform negatives into positives. Existential positive psychology considers the transformation of suffering as the new holy grail to wellbeing, teaching us how to accept and transcend negative emotions and experiences so that we can become better and stronger (Bonanno, 2021; Targ & Hurtak, 2006; Wong & Worth, 2017).

The Thread of Self-Transcendence

A central theme of existential positive psychology is self-transcendence (Wong, Arslan, et al., 2021). Self-transcendence is a uniquely human phenomenon that involves a fundamental reorientation from an egotistic focus to a selflessness emphasis on something outside or greater than oneself (Kaufman, 2020; Yaden et al., 2017). This was eloquently described by Viktor Frankl in his book *Man's Search for Meaning* (1946/1985): "The more one forgets himself—by giving himself to a cause to serve or another person to love—the more human he is and the more he actualizes himself… In other words, self-actualization is possible only as a side-effect of self-transcendence" (p. 133).

The concept of self-transcendence is essentially a paradoxical approach to personal growth and self-actualization. From the perspective of existential positive psychology, self-transcendence is necessary because we can only become our best selves through two interrelated processes: (a) sinking our roots into the dark soil of suffering and transcending obstacles and inherent limitations while simultaneously (b) striving toward our highest ideals of serving the greater good (Reischer et al., 2020; Wong, 2021). Engaging both processes is challenging, which is why the path of self-transcendence is narrow and long. The first of the interrelated processes can be particularly intimidating because it requires that we overcome our deep-seated desire to avoid suffering and instead make the choice to courageously encounter our suffering (Wong, 2017). Yet, there are many inspiring examples of how this approach to suffering can be adaptive and contribute to wellbeing. For example, reflecting on the myth of Sisyphus who was condemned to a futile, lifelong struggle of pushing a large boulder up a mountain each day, Camus (1942/1955) concluded that "the struggle itself toward the heights is enough to fill a man's heart. One must imagine Sisyphus happy" (p. 123). Similarly, Frankl (1946/1985) wrote that our capacity to accept our fate and the suffering that comes along with it provides us with an opportunity to deepen our sense of meaning in life. If we can learn to embrace our suffering, even circumstances that might ordinarily lead to an "existential vacuum" (Frankl, 1946/1985), we can meet our suffering in a way that enables us to transform it so that we become better rather than bitter. Ultimately, one's wellbeing, and indeed one's destiny, depends on how one responds to the trials and tribulations that we all go through at some point in our lives.

Implications for Research and Practice

In this chapter, we proposed a paradigm shift toward existential positive psychology, which offers an integrative framework that recognizes the dialectical interaction between the dark (negative psychology) and bright (positive psychology) sides of human experience. The existential positive psychology paradigm not only enables us to see life as it is, but it also has heuristic value in advancing research and practice dedicated to the promotion of wellbeing. We highlight some of these potential implications below.

A key premise of the existential positive psychology perspective is that sustainable wellbeing requires both actions focused on enhancing happiness and actions oriented toward the transformation of suffering. In contrast with common conceptions of happiness as a transient positive emotional state, existential positive psychology emphasizes mature happiness as a distinct dimension of happiness that is characterized by durable contentment or inner peace (Wong & Bowers, 2019). Mature happiness tends to be more durable because it is based on the human capacity to achieve balance and self-transcendence (Dambrun et al., 2012), which should make it more resistant to the disruptive impact that negative circumstances can have on wellbeing. Although a growing body of literature supports the benefits of mature happiness for a flourishing life (Bowers, 2020; Chen et al., 2021), research is needed to explore the unique contributions of both mature happiness and transformed suffering to sustainable wellbeing. Some theorists have emphasized that humans are hardwired to survive and transcend hardships (Frankl, 1946/1985; Wong, 2021), and there is evidence suggesting that stress can contribute to personal growth and transformation if we learn how to harness it as a source of energy (McGonigal, 2015). However, little is known about whether suffering that is positively transformed into an asset contributes to wellbeing beyond mature happiness (and other dimensions of happiness), which could shed important light on how people might be able to develop and grow from suffering in ways that ultimately support their wellbeing.

Rigorous, ground-breaking research will be necessary for generating evidence that captures the breadth and depth of how transformed suffering might promote wellbeing. For example, psychobiography could provide opportunities to acquire an in-depth, holistic understanding of the complex and dynamic processes involved in transforming suffering across the lifespan (Mayer & May, 2019). On the other hand, well-designed longitudinal studies will be important for exploring the effects of transformed suffering on outcomes over long periods of time, which may be especially necessary because our capacities to positively transform suffering can take time to develop. Research along these lines will require new measures to assess capacities that might enable people to transform and transcend suffering, as demonstrated by recently developed measures of tragic optimism (Leung et al., 2021) and existential gratitude (Jans-Beken & Wong, 2021). Mixed methods approaches that balance methodological rigor needed for estimating causality with rich insight acquired through personal narratives offer a particularly promising avenue for uncovering the complexities involved in transforming suffering and its implications for wellbeing.

Innovative and engaging interventions are needed to help people enhance mature happiness and positively transform suffering. Many therapeutic models that have roots in existential positive psychology could be used to inform the development of such interventions (e.g., integrative meaning therapy; Wong, 2016), but special consideration should be given to formulating impactful evidence-based interventions (e.g., self-guided psychoeducational approaches) that have the potential to be disseminated widely and can be made available to the public at low cost. Although these kinds of interventions would not represent an alternative to psychotherapy, they could be beneficial to many people who are not flourishing or wish to flourish more fully.

Conclusion

In a world that is constantly changing and at times can be difficult to make sense of, we need a positive psychology that addresses the complexity of the human condition and offers an integrative framework that lays a firm foundation for future theory, research, and practice that can support the needs of humanity. Toward this end, we explored existential positive psychology as a well-rounded approach that addresses the dialectical interplay between the dark and bright sides of life for a more holistic understanding of wellbeing. Rather than neglect or underemphasize suffering, existential positive psychology posits that sustainable wellbeing necessarily depends on our capacity to embrace and transcend suffering. It implicitly recognizes that suffering adds depth to our lives by teaching us who we really are and sets the stage for us to develop inner resources and virtues that enable us to transcend life's challenges. Acknowledging the centrality of suffering to wellbeing represents a key paradigm shift in positive psychology, and the integrative approach that existential positive psychology provides could be an important next step on our journey to building a richer and more refined understanding of how individuals and society can fully flourish.

References

Bland, A. M. (2020). Existential givens in the COVID-19 crisis. *Journal of Humanistic Psychology, 60*(5), 710–724. https://doi.org/10.1177/0022167820940186

Bonanno, G. A. (2021). *The end of trauma: How the new science of resilience is changing how we think about PTSD*. Basic Books.

Bowers, V. L. (2020). Transpersonal psychology and mature happiness in the context of counseling. *Counselling Psychology Quarterly, 33*(4), 572–582. https://doi.org/10.1080/09515070.2019.1634518

Camus, A. (1942/1955). *The myth of Sisyphus and other essays (Trans. J. O'Brien)*. Vintage Books.

Chen, C.-H., Chang, S.-M., & Wu, H.-M. (2021). Discovering and approaching mature happiness: The implementation of the CasMac model in a university English class. *Frontiers in Education, 6*, 648311. https://doi.org/10.3389/feduc.2021.648311

Counted, V., Cowden, R. G., & Ramkissoon, H. (2021). *Place and post-pandemic flourishing: Disruption, adjustment, and healthy behaviors*. Springer. https://doi.org/10.1007/978-3-030-82580-5

Cowden, R. G., Counted, V., & Ramkissoon, H. (2021). Place attachment and suffering during a pandemic. In V. Counted, R. G. Cowden, & H. Ramkissoon, *Place and Post-pandemic Flourishing: Disruption, Adjustment, and Healthy Behaviors* (pp. 45–54). Springer. https://doi.org/10.1007/978-3-030-82580-5_4

Cowden, R. G., Davis, E. B., Counted, V., Chen, Y., Rueger, S. Y., VanderWeele, T. J., Lemke, A. W., Glowiak, K. J., & Worthington, E. L., Jr. (2021). Suffering, mental health, and psychological well-being during the COVID-19 pandemic:

A longitudinal study of U.S. adults with chronic health conditions. *Wellbeing, Space and Society, 2*, 100048. https://doi.org/10.1016/j.wss.2021.100048

Cowden, R. G., Rueger, S. Y., Davis, E. B., Counted, V., Kent, B. V., Chen, Y., VanderWeele, T. J., Rim, M., Lemke, A. W., & Worthington, E. L., Jr. (2021). Resource loss, positive religious coping, and suffering during the COVID-19 pandemic: A prospective cohort study of US adults with chronic illness. *Mental Health, Religion & Culture, 25*, 288–304. https://doi.org/10.1080/13674676.2021.1948000

Coyne, J. C., & Tennen, H. (2010). Positive psychology in cancer care: Bad science, exaggerated claims, and unproven medicine. *Annals of Behavioral Medicine, 39*(1), 16–26. https://doi.org/10.1007/s12160-009-9154-z

Dambrun, M., Ricard, M., Després, G., Drelon, E., Gibelin, E., Gibelin, M., Loubeyre, M., Py, D., Delpy, A., Garibbo, C., Bray, E., Lac, G., & Michaux, O. (2012). Measuring happiness: From fluctuating happiness to authentic–durable happiness. *Frontiers in Psychology, 3*, 16. https://doi.org/10.3389/fpsyg.2012.00016

Deng, K., Wong, Y. J., Li, J. P. F., & McCullough, K. M. (2020). Dialectical coping and well-being among Chinese college students: The mediating role of resilience. *Counselling Psychology Quarterly, 35*, 363–380. https://doi.org/10.1080/09515070.2020.1783641

De Kock, J. H., Latham, H. A., Cowden, R. G., Cullen, B., Narzisi, K., Jerdan, S., Muñoz, S.-A., Leslie, S. J., McNamara, N., Boggon, A., & Humphry, R. W. (2022). The mental health of NHS staff during the COVID-19 pandemic: Two-wave Scottish cohort study. *British Journal of Psychiatry Open, 8*(1), E23. https://doi.org/10.1192/bjo.2021.1079

Fowers, B. J., Richardson, F. C., & Slife, B. D. (2017). *Frailty, suffering, and vice: Flourishing in the face of human limitations*. American Psychological Association.

Frankl, V. E. (1946/1985). *Man's search for meaning*. Simon & Schuster.

Friedman, H. L., & Robbins, B. D. (2012). The negative shadow cast by positive psychology: Contrasting views and implications of humanistic and positive psychology on resiliency. *The Humanistic Psychologist, 40*, 87–102. https://doi.org/10.1080/08873267.2012.643720

Gable, S. L., & Haidt, J. (2005). What (and why) is positive psychology? *Review of General Psychology, 9*(2), 103–110. https://doi.org/10.1037/1089-2680.9.2.103

Govender, K., Cowden, R. G., Nyamaruze, P., Armstrong, R. M., & Hatane, L. (2020). Beyond the disease: Contextualized implications of the COVID-19 pandemic for children and young people living in eastern and southern Africa. *Frontiers in Public Health, 8*, 504. https://doi.org/10.3389/fpubh.2020.00504

Gruber, J., Mauss, I. B., & Tamir, M. (2011). A dark side of happiness? How, when, and why happiness is not always good. *Perspectives on Psychological Science, 6*(3), 222–233. https://doi.org/10.1177/1745691611406927

Held, B. S. (2004). The negative side of positive psychology. *Journal of Humanistic Psychology, 44*(1), 9–46. https://doi.org/10.1177/0022167803259645
Hershfield, H. E., Scheibe, S., Sims, T. L., & Carstensen, L. L. (2013). When feeling bad can be good: Mixed emotions benefit physical health across adulthood. *Social Psychological and Personality Science, 4*(1), 54–61. https://doi.org/10.1177/1948550612444616
Ho, S., Cook, K. V., Chen, Z. J., Kurniati, N. M. T., Suwartono, C., Widyarini, N., Wong, P. T. P., & Cowden, R. G. (2022). Suffering, psychological distress, and well-being in Indonesia: A prospective cohort study. *Stress and Health*. Advance online publication. https://doi.org/10.1002/smi.3139
Jacobi, C. J., Cowden, R. G., & Vaidyanathan, B. (2022). Associations of changes in religiosity with flourishing during the COVID-19 pandemic: A study of faith communities in the United States. *Frontiers in Psychology, 13*, 805785. https://doi.org/10.3389/fpsyg.2022.805785
Jans-Beken, L., & Wong, P. T. P. (2021). Development and preliminary validation of the Existential Gratitude Scale (EGS). *Counselling Psychology Quarterly, 34*(1), 72–86. https://doi.org/10.1080/09515070.2019.1656054
Kashdan, T., & Biswas-Diener, R. (2014). *The upside of your dark side: Why being your whole self–not just your "good" self–drives success and fulfillment*. Avery.
Kaufman, S. B. (2020). *Transcend: The new science of self-actualization*. Tarcher Perigee.
Kemp, A. H., & Fisher, Z. (2022). Wellbeing, whole health and societal transformation: Theoretical insights and practical applications. *Global Advances in Health and Medicine, 11*, 1–16. https://doi.org/10.1177/21649561211073077
Kjell, O. N. E. (2011). Sustainable well-being: A potential synergy between sustainability and well-being research. *Review of General Psychology, 15*(3), 255–266. https://doi.org/10.1037/a0024603
Kuhn, T. S. (1962). *The structure of scientific revolutions*. University of Chicago Press.
Lazarus, R. S. (2003). Does the positive psychology movement have legs? *Psychological Inquiry, 14*(2), 93–109. https://doi.org/10.1207/S15327965PLI1402_02
Lee, M. T., Kubzansky, L. D., & VanderWeele, T. J. (Eds.). (2021). *Measuring well-being: Interdisciplinary perspectives from the social sciences and the humanities*. Oxford University Press. https://doi.org/10.1093/oso/9780197512531.001.0001
Leung, M. M., Arslan, G., & Wong, P. T. P. (2021). Tragic optimism as a buffer against COVID-19 suffering and the psychometric properties of a brief version of the life attitudes scale. *Frontiers in Psychology, 12*, 646843. https://doi.org/10.3389/fpsyg.2021.646843

Lomas, T., & Ivtzan, I. (2016). Second wave positive psychology: Exploring the positive-negative dialectics of wellbeing. *Journal of Happiness Studies, 17*(4), 1753–1768. https://doi.org/10.1007/s10902-015-9668-y

Lomas, T., Waters, L., Williams, P., Oades, L. G., & Kern, M. L. (2021). Third wave positive psychology: Broadening towards complexity. *The Journal of Positive Psychology, 16*(5), 660–674. https://doi.org/10.1080/17439760.2020.1805501

Mayer, C.-H. (2020). Transforming shame in the workplace, leadership and organisation: Contributions of positive psychology movements to the discourse. In S. Dhiman & J. Marquis (Eds.), *New Horizons in Positive Leadership and Change* (pp. 313–331). Springer.

Mayer, C.-H. (2021). Albert Camus—A psychobiographical approach in times of Covid-19. *Frontiers in Psychology, 12*, 644579. https://doi.org/10.3389/fpsyg.2021.644579

Mayer, C.-H., & May, M. (2019). The positive psychology movements PP1.0 and PP2.0 in psychobiography. In C.-H. Mayer & Z. Kovary (Eds.), *New Trends in Psychobiography* (pp. 155–171). Springer.

Mayer, C.-H., & Vanderheiden, E. (Eds.). (2019). *The bright side of shame. Transforming and growing through practical applications in cultural contexts.* Springer.

Mayer, C.-H., & Vanderheiden, E. (2021). Transforming shame in the pandemic: An international study. *Frontiers in Psychology, 12*, 641076. https://doi.org/10.3389/fpsyg.2021.641076

McGonigal, K. (2015). *The upside of stress: Why stress is good for you, and how to get good at it.* Avery.

Miller, A. (2008). A critique of positive psychology—Or 'the new science of happiness'. *Journal of Philosophy of Education, 42*, 591–608. https://doi.org/10.1111/j.1467-9752.2008.00646.x

Reischer, H. N., Roth, L. J., Villarreal, J. A., & McAdams, D. P. (2020). Self-transcendence and life stories of humanistic growth among late-midlife adults. *Journal of Personality, 89*(2), 305–324. https://doi.org/10.1111/jopy.12583

Ryff, C. D. (2022). Positive psychology: Looking back and looking forward. *Frontiers in Psychology, 13*, 840062. https://doi.org/10.3389/fpsyg.2022.840062

Seligman, M. E. P. (2011). *Flourish: A visionary new understanding of happiness and well-being.* Free Press.

Shiba, K., Cowden, R. G., Counted, V., VanderWeele, T. J., & Fancourt, D. (2022). Associations of home confinement during COVID-19 lockdown with subsequent health and well-being among UK adults. *Current Psychology.* Advance online publication. https://doi.org/10.1007/s12144-022-03001-5

Shiba, K., Cowden, R. G., Gonzalez, N., Lee, M. T., Lomas, T., Lai, A. Y., & VanderWeele, T. J. (2022). Global trends of mean and inequality in multidimensional wellbeing: Analysis of 1.2 million individuals from 162 countries, 2009–2019. *Frontiers in Public Health, 10*, 824960. https://doi.org/10.3389/fpubh.2022.824960

Targ, R., & Hurtak, J. J. (2006). *The end of suffering: Fearless living in troubled times…or, how to get out of hell free*. Hampton Roads Publishing.

VanderWeele, T. J. (2017). On the promotion of human flourishing. *Proceedings of the National Academy of Sciences, 114*(31), 8148–8156. https://doi.org/10.1073/pnas.1702996114

Wong, P. T. P. (2011). Positive psychology 2.0: Towards a balanced interactive model of the good life. *Canadian Psychology, 52*(2), 69–81. https://doi.org/10.1037/a0022511

Wong, P. T. P. (2012). Toward a dual-systems model of what makes life worth living. In P. T. P. Wong (Ed.), *The Human Quest for Meaning: Theories, Research, and Applications* (2nd ed., pp. 3–22). Routledge.

Wong, P. T. P. (2016). Integrative meaning therapy: From logotherapy to existential positive interventions. In P. Russo-Netzer, S. E. Schulenberg, & A. Batthyany (Eds.), *Clinical Perspectives on Meaning: Positive and Existential Psychotherapy* (pp. 323–342). Springer. https://doi.org/10.1007/978-3-319-41397-6_16

Wong, P. T. P. (2017). Meaning-centered approach to research and therapy, second wave positive psychology, and the future of humanistic psychology. *The Humanistic Psychologist, 45*(3), 207–216. https://doi.org/10.1037/hum0000062

Wong, P. T. P. (2019). Second wave positive psychology's (PP 2.0) contribution to counselling psychology. *Counselling Psychology Quarterly, 32*(3–4), 275–284. https://doi.org/10.1080/09515070.2019.1671320

Wong, P. T. P. (2020). *Made for resilience and happiness: Effective coping with COVID-19 according to Viktor E. Frankl and Paul T. P. Wong*. INPM Press.

Wong, P. T. P. (2021). What is existential positive psychology (PP 2.0)? Why is it necessary for mental health during the pandemic? *International Journal of Existential Positive Psychology, 10*(1), 1–16. https://www.meaning.ca/ijepp-article/vol10-no1/what-is-existential-positive-psychology-pp-2-0-why-is-it-necessary-for-mental-health-during-the-pandemic/

Wong, P. T. P., Arslan, G., Bowers, V. L., Peacock, E. J., Kjell, O. N. E., Ivtzan, I., & Lomas, T. (2021). Self-transcendence as a buffer against COVID-19 suffering: The development and validation of the self-transcendence measure-B. *Frontiers in Psychology, 12*, 4229. https://doi.org/10.3389/fpsyg.2021.648549

Wong, P. T. P., & Bowers, V. (2019). Mature happiness and global wellbeing in difficult times. In N. R. Silton (Ed.), *Scientific Concepts Behind Happiness, Kindness, and Empathy in Contemporary Society* (pp. 112–134). IGI Global.

Wong, P. T. P., Mayer, C.-H., & Arslan, G. (2021). Editorial: COVID-19 and existential positive psychology (PP2.0): The new science of self-transcendence. *Frontiers in Psychology, 12*, 800308. https://doi.org/10.3389/fpsyg.2021.800308

Wong, P. T. P., & Worth, P. (2017). The deep-and-wide hypothesis in giftedness and creativity. *Psychology and Education, 54*(3/4), 11–23.

Wong, P. T. P., & Yu, T. T. F. (2021). Existential suffering in palliative care: An existential positive psychology perspective. *Medicina, 57*(9), 924. https://doi.org/10.3390/medicina57090924

Yaden, D. B., Haidt, J., Hood, R. W., Jr., Vago, D. R., & Newberg, A. B. (2017). The varieties of self-transcendent experience. *Review of General Psychology, 21*(2), 143–160. https://doi.org/10.1037/gpr0000102

Yalom, I. D. (1980). *Existential psychotherapy*. Basic Books.

Zerwas, F. K., & Ford, B. Q. (2021). The paradox of pursuing happiness. *Current Opinion in Behavioral Sciences, 39*, 106–112. https://doi.org/10.1016/j.cobeha.2021.03.006

CHAPTER 3

Beyond Us: Building Collective Wellbeing

Louise Lambert, Meg A. Warren, Rachael J. Waldrop, Sabrina Al Subaiei, Alecia Eubanks, and Jen L. Fisher

Abstract The mission of positive psychology is to explore and build wellbeing in the lives of individuals the world over; however, like all theoretical frameworks, its evolution propels researchers and practitioners into increasingly complex issues. Developments in the field promote a broader view on wellbeing, focusing on culture, religion, social interactions and climate action. Global issues have encouraged the need for a systemic view of wellbeing that is beyond the remit of individuals and extends to organizations,

---

The original version of this chapter was revised. Incorrect spelling of the author's name Rachel J. Waldrop has been corrected to Rachael J. Waldrop. A correction to this chapter can be found at https://doi.org/10.1007/978-3-031-18329-4_12

L. Lambert (✉)
Department of Psychology, Canadian University Dubai,
Dubai, United Arab Emirates

M. A. Warren
Western Washington University, Bellingham, WA, USA
e-mail: Meg.Warren@wwu.edu

R. J. Waldrop
Kent School District, Kent, WA, USA

A. H. Kemp, D. J. Edwards (eds.), *Broadening the Scope of Wellbeing Science*, https://doi.org/10.1007/978-3-031-18329-4_3

governments and societies alike. In short, the time has come for a global, contextual and systemic understanding of wellbeing. Here, we address the development of these frameworks and three emerging areas as examples of their utility; namely, climate change, egalitarianism as an aid to intergroup relations, and the need for purposeful non-Western views of wellbeing.

Keywords Wellbeing • Culture • Egalitarianism • Climate change • Positive psychology

Positive psychology's (PP) beginnings built the foundations for the development of theories upon which to grow its science (Lomas & Ivtzan, 2016; Lomas et al., 2020). A second wave of PP (Lomas & Ivtzan, 2016; Wong, 2011) addressed the positivity imbalance, included non-Western voices and developed means to improve wellbeing via positive psychology interventions (PPIs; Parks & Biswas-Diener, 2013). Recently, a third iteration (PP 3.0) elevates inquiry beyond individuals, highlighting the role of context, sociocultural factors, systems and groups to address issues that influence wellbeing (Kern et al., 2020; Lomas et al., 2020; Mead et al., 2019; Warren & Bordoloi, 2020). These models posit that wellbeing is not possible via individual intervention alone; they construe wellbeing as a complex system in which the relationships individuals share with their communities and environments serve as the nexus in which wellbeing evolves (Mead et al., 2021).

Accordingly, such models are interdisciplinary. For instance, Kemp et al.'s (2017) GENIAL (Genomics—Environment—vagus Nerve—social Interaction—Allostatic regulation—Longevity) model integrates evidence from PP and other disciplines including psychophysiology and epidemiology to illustrate the psychological, physiological and sociocultural mechanisms underlying health and wellbeing. Similarly, Systems Informed Positive Psychology (SIPP; Kern et al., 2020) explores the dynamism between individuals and their political, social, cultural, ecological systems to develop new understandings and interventions. Such models advocate for more plurality, unearthing non-Western conceptualizations of wellbeing (e.g., Lambert, Lomas et al., 2020). Relying on diverse research

S. Al Subaiei
United Arab Emirates University, Abu Dhabi, United Arab Emirates

A. Eubanks • J. L. Fisher
International Psychology, The Chicago School of Professional Psychology, Chicago, IL, USA

methods, the goal is to integrate and recontextualize existing knowledge and generate new ideas and theories (Mead et al., 2021), the end-goal being a transdisciplinary science of wellbeing. This broader view of wellbeing can mobilize individuals around higher-order goals that generate greater collective wellbeing going forward (Waters et al., 2021).

Below are examples of inquiry that illustrate systems thinking and serve to understand and build greater forms of collective wellbeing. They include: (1) expanding the notion of wellbeing to include social, cultural and religious issues, (2) nurturing egalitarianism as an aid to intergroup relations, and (3) positioning climate action as critical to human wellbeing.

Post-COVID-19: Looking to Context

Positive psychology interventions (PPIs) are strategies designed to tackle depression (Seligman et al., 2005). They have been effectively used (Carr et al., 2020; Hendriks et al., 2020), including during the pandemic (Brouzos et al., 2021; Yaden et al., 2021) where levels of anxiety and depression rose dramatically (Castaldelli-Maia et al., 2021; Gloster et al., 2020). Inequalities also grew (Perry et al., 2021; Slobodin & Cohen, 2020; Warren & Bordoloi, 2020). Lower socioeconomic status undermined access to healthcare which worsened infection rates (Gloster et al., 2020; Slobodin & Cohen, 2020). Prejudice, discrimination, marginalization and mistrust in authority further fuelled inequalities, including cultural and religious strife (Schleussner et al., 2016; Slobodin & Cohen, 2020). The pandemic's impact also stoked extremism across Western nations (Marone, 2021), anti-science rhetoric (Sanchez & Dunning, 2021), and political polarization (Gonsalves & Yamey, 2020). PP studies have addressed intellectual humility (Hodge et al., 2020), and feminine leadership during COVID-19, with its focus on citizen responsibility and human compassion (Mayer & May, 2021) as examples of work mitigating these issues.

Indeed, Lambert, Warren et al. (2020) have called for a PPI expansion to address system interactions that give rise to inequalities and perpetuate exclusion. For instance, PPIs are often criticized for their Western assumptions, explicitly unstated when used in non-Western contexts (i.e., Hendriks et al., 2018; Kim et al., 2018). The dominant view of wellbeing itself reflects a "WEIRD" worldview, i.e., Western, educated, industrialized, rich, democratic (Hendriks et al., 2019); yet, the majority of nations in which PPIs are now used are collectivistic and religious, factors that largely remain unaccounted for in PPI use, but which lead to wellbeing (Hatala et al., 2016; Lucette et al., 2016; Psaila, 2019).

The pandemic has redirected attention to inequality (Paremoer et al., 2021), as few events have been as potent in showing the interconnections and impacts of systems on individuals. It has forced researchers to look beyond individuals and attend to the wellbeing of societies overall. Exploring the interactions between individuals and their systems; their relationships with one another; the effects of cultural, social, political and religious views of the "other"; as well as on wellbeing, points the way forward (Lambert, Lomas et al., 2020).

Egalitarianism: Supporting Social Wellbeing

Individuals form groups in which they build collective identities (Tajfel, 1981); yet, they show hostility, prejudice and ostracism toward outgroups (Brewer, 1999) despite social values against these (Crandall et al., 2002). While gender and race biases have decreased in time (Charlesworth & Banaji, 2019), actions toward redressing inequalities have often been perceived as superficial (Anderson & McClain, 2020), and inequality persists (Perry et al., 2021; Warren & Bordoloi, 2020). While social psychology focuses on eliminating such responses (Mallett & Monteith, 2019; Paluck et al., 2020), PP has remained silent on intergroup and intercultural dynamics (Rao & Donaldson, 2015), and avoided prescribing how we ought to relate. As sociocultural contexts become more diverse and PP 3.0 incorporates contextual issues, an investigation of the positively oriented principles underlying pro-intergroup behavior is needed. Egalitarianism can be a guiding value that motivates the inclusive treatment of marginalized peoples.

Many factors work against successful intergroup interaction, such as prejudice, implicit bias, microaggressions and stereotype threat, with much research addressing these (Ashburn-Nardo et al., 2008; Cox & Devine, 2019; Devine et al., 2017; Lai et al., 2013). Underlying many prejudice reduction strategies include avoidance-oriented values via advancement of end-goals like nonprejudice and antiracism. Internalized through social norms, interactions or organizational policies (Oyserman, 2015), they dominate the intergroup literature; yet, studies show that behavior guided by avoidant values and goals (e.g., to not discriminate) alone is experienced as impersonal (Plant & Butz, 2006; Plant et al., 2010) and leaves individuals with fewer cognitive resources to intentionally develop positive relationships (Trawalter & Richeson, 2006; Zabel et al., 2015).

A PP 3.0 framework suggests that approach-oriented rather than avoidance-oriented goals may lead to greater intergroup wellbeing. Values rooted in prevention (i.e., nonprejudiced, antiracism) inhibit and prompt the restraint of behaviors that allow for the avoidance of actions that create psychological and social harm. Alternatively, values rooted in promotion (i.e., justice, egalitarianism) stimulate "doing good" and encourage helping, socially minded action, and life satisfaction (Higgins, 1997; Janoff-Bulman & Carnes, 2013; Meyer & Warren, 2021; Schutte & Malouff, 2019). Solely focusing on prevention goals helps avoid displaying prejudice but produces little change, while promotion goals may result in friendlier interaction but inadequately reshape bias. Because individuals are influenced by their own experiences and values, as well as the actions and collective values being promoted (Kern et al., 2020), prevention and promotion goals must be balanced.

By shaping inquiry on the antecedents of egalitarian behavior, positive intergroup relational processes and intergroup outcomes, a PP lens reveals a meaningful, relationship-building construct. As PP lays the groundwork for investigating the mechanisms that guide positive outcomes—i.e., the Values in Action classification which identifies strengths that positively influence environments (Peterson & Seligman, 2004)—understanding progressive, justice-centered values that encourage proactive and preventative strategies is possible. Yet, this demands PP be value-clear and expressly promote a vision of wellbeing that nurtures belonging and corrects harm. Positively oriented values that stimulate and promote friendly or prosocial behavior, and intentional, authentic allyship (Warren et al., 2022) can help societies navigate intergroup interactions, bolster organizational culture and policy (Pasha-Zaidi et al., 2021), and shape norms that guide how individuals perceive and care for one another.

Climate Action: Wellbeing Demands It

The effects of climate change on individuals and societies are far reaching; pandemics are attributed in part, to the close proximity and encroachment on wildlife due to human activity (Lindahl & Grace, 2015). Climate change destabilizes food security and affects human migration, economic hardship and war (Abel et al., 2019; Agnew, 2012; Campbell-Lendrum et al., 2015; Gleick, 2014). For every standard deviation of increased temperature or extreme rainfall, a 4–14% increase in conflict (Hsiang et al., 2013), violence and shootings (Anderson, 2001; Towers et al., 2018) have been recorded. The impact of extreme weather on post-traumatic stress disorder, depression, anxiety, substance abuse and suicides is well-documented (Burke et al., 2018; Clayton, 2020; Hayes et al., 2018; Lai et al., 2015; Liu et al., 2020; Manning & Clayton, 2018), and climate anxiety is growing (Clayton, 2020).

It is not only in its absence that nature matters, but also in its presence. In densely natured residential areas, studies show less psychological distress (Astell-Burt & Feng, 2019; Taylor et al., 2015), fewer antidepressant prescriptions, and more physical activity (de Bell et al., 2020; Koohsari et al., 2015) and sleep (Grigsby-Toussaint et al., 2015). Associated with less crime and more prosocial behavior (Bogar & Beyer, 2016; Putra et al., 2020; Zelenski et al., 2015), time in nature also leads to higher subjective wellbeing and positive emotional experiences (Passmore & Holder, 2017; Ward Thompson et al., 2016; White et al., 2019). Yet, individuals living in urban spaces may be less likely to appreciate nature or engage in pro-environmental actions as they are less exposed and connected to nature (Alcock et al., 2020; Weinstein et al., 2015; Zelenski et al., 2015).

As nature diminishes, transformations await (Cianconi et al., 2020); but inequality is certain as nations vary in their ability to invest in green technologies, rebuild infrastructure, or prevent adverse fallouts of climate change through political will, policy implementation, and identification of risk. Government action as well as individual and collective behavior is needed both to slow down climate change and to live with it. Reducing consumption, adopting a circular economy, or collective housing, rest on individual choices. Using PP to motivate families to have fewer children and increase subjective wellbeing is an example of such work (Tal & Kerret, 2020), while exploring how social values are constructed in the face of sustainability challenges is another (Raymond & Raymond, 2019). As our cultural and social lives, as well as economies and livelihoods depend on the natural world, an ecological approach to wellbeing is imperative.

Conclusion

The world is small: social movements, climate change and religious/cultural factors affect everyone. PP can no longer afford to focus on individuals alone, as wellbeing is influenced by systemic factors beyond their grasp. Science can help us understand what moves people to action and how to foster communal responsibility, especially in light of denial and complacency. We challenge PP adherents to promote the notion of well-being beyond psychological boundaries, as our very survival is affected by, and bound to, it.

References

Abel, G. J., Brottrager, M., Crespo Cuaresma, J., & Muttarak, R. (2019). Climate, conflict and forced migration. *Global Environmental Change, 54*, 239–249.

Agnew, R. (2012). Dire forecast: A theoretical model of the impact of climate change on crime. *Theoretical Criminology, 16*(1), 21–42.

Alcock, I., White, M. P., Pahl, S., Duarte-Davidson, R., & Fleming, L. E. (2020). Associations between pro-environmental behaviour and neighbourhood nature, nature visit frequency and nature appreciation: Evidence from a nationally representative survey in England. *Environment International, 136*, 105441. https://doi.org/10.1016/j.envint.2019.105441

Anderson, C. A. (2001). Heat and violence. *Current Directions in Psychological Science, 10*(1), 33–38. https://doi.org/10.1111/1467-8721.00109

Anderson, M., & McClain, C. (2020, August 12). Americans see pressure, rather than genuine concern, as big factor in company statements about racism. *Pew Research Center*. https://www.pewresearch.org/facttank/2020/08/12/americans-see-pressure-rather-than-genuine-concern-as-big-factor-in-company-statements-aboutracism/

Ashburn-Nardo, L., Morris, K. A., & Goodwin, S. A. (2008). The confronting prejudiced responses (CPR) model: Applying CPR in organizations. *Academy of Management Learning & Education, 7*(3), 332–342. https://doi.org/10.5465/amle.2008.34251671

Astell-Burt, T., & Feng, X. (2019). Association of urban green space with mental health and general health among adults in Australia. *JAMA Network Open, 2*(7), e198209. https://doi.org/10.1001/jamanetworkopen.2019.8209

Bogar, S., & Beyer, K. M. (2016). Green space, violence, and crime: A systematic review. *Trauma, Violence & Abuse, 17*(2), 160–171. https://doi.org/10.1177/1524838015576412

Brewer, M. B. (1999). The psychology of prejudice: Ingroup love or outgroup hate? *Journal of Social Issues, 55*, 429–444.

Brouzos, A., Vassilopoulos, S. P., Baourda, V. C., Tassi, C., Stavrou, V., Moschou, K., & Brouzou, K. O. (2021). "staying home – feeling positive": Effectiveness of an on-line positive psychology group intervention during the covid-19 pandemic. Current Psychology, Advance online publication. https://doi.org/10.1007/s12144-021-01613-x

Burke, M., González, F., Baylis, P., Heft-Neal, S., Baysan, C., Basu, S., & Hsiang, S. (2018). Higher temperatures increase suicide rates in the United States and Mexico. *Nature Climate Change, 8*(8), 723–729.

Campbell-Lendrum, D., Manga, L., Bagayoko, M., & Sommerfeld, J. (2015). Climate change and vector-borne diseases: What are the implications for public health research and policy? *Philosophical Transactions of the Royal Society of London Series B, Biological Sciences, 370*(1665), 20130552.

Carr, A., Cullen, K., Keeney, C., Canning, C., Mooney, O., Chinseallaigh, E., et al. (2020). Effectiveness of positive psychology interventions: A systematic review and meta-analysis. *The Journal of Positive Psychology, 49*(10), 1943–1960. https://doi.org/10.1080/17439760.2020.1818807

Castaldelli-Maia, J. M., Marziali, M. E., Lu, Z., & Martins, S. S. (2021). Investigating the effect of national government physical distancing measures on depression and anxiety during the COVID-19 pandemic through meta-analysis and meta-regression. *Psychological Medicine, 51*(6), 881–893. https://doi.org/10.1017/S0033291721000933

Charlesworth, T. E., & Banaji, M. R. (2019). Patterns of implicit and explicit attitudes: I. Long-term change and stability from 2007 to 2016. *Psychological Science, 30*(2), 174–192. https://doi.org/10.1177/0956797618813087

Cianconi, P., Betrò, S., & Janiri, L. (2020). The impact of climate change on mental health: A systematic descriptive review. *Frontiers in Psychiatry, 11*, 74. https://doi.org/10.3389/fpsyt.2020.00074

Clayton, S. (2020). Climate anxiety: Psychological responses to climate change. *Journal of Anxiety Disorders, 74*, 102263.

Cox, W. T., & Devine, P. G. (2019). The prejudice habit-breaking intervention: An empowerment-based confrontation approach. In R. K. Mallet & M. J. Monteith (Eds.), *Confronting Prejudice and Discrimination* (pp. 249–274). Academic Press. https://doi.org/10.1016/B978-0-12-814715-3.00015-1

Crandall, C. S., Eshleman, A., & O'Brien, L. (2002). Social norms and the expression and suppression of prejudice: The struggle for internalization. *Journal of Personality and Social Psychology, 82*(3), 359–378. https://doi.org/10.1037/0022-3514.82.3.359

de Bell, S., White, M., Griffiths, A., Darlow, A., Taylor, T., Wheeler, B., et al. (2020). Spending time in the garden is positively associated with health and wellbeing: Results from a national survey in England. *Landscape and Urban Planning, 200*, 103836. https://doi.org/10.1016/j.landurbplan.2020.103836

Devine, P. G., Forscher, P. S., Cox, W. T. L., Kaatz, A., Sheridan, J., & Carnes, M. (2017). A gender bias habit-breaking intervention led to increased hiring of female faculty in STEMM departments. *Journal of Experimental Social Psychology, 73*, 211–215. https://doi.org/10.1016/j.jesp.2017.07.002

Gleick, P. H. (2014). Water, drought, climate change, and conflict in Syria. *Weather, Climate, and Society, 6*, 331–340.

Gloster, A. T., Lamnisos, D., Lubenko, J., Presti, G., Squatrito, V., Constantinou, M., et al. (2020). Impact of COVID-19 pandemic on mental health: An international study. *PLoS One, 15*, e0244809. https://doi.org/10.1371/journal.pone.0244809

Gonsalves, G., & Yamey, G. (2020). Political interference in public health science during COVID-19. *British Medical Journal, 371*, m3878. https://doi.org/10.1136/bmj.m3878

Grigsby-Toussaint, D. S., Turi, K. N., Krupa, M., Williams, N. J., Pandi-Perumal, S. R., & Jean-Louis, G. (2015). Sleep insufficiency and the natural environment: Results from the US behavioral risk factor surveillance system survey. *Preventive Medicine, 78*, 78–84. https://doi.org/10.1016/j.ypmed.2015.07.011

Hatala, A. R., Desjardins, M., & Bombay, A. (2016). Reframing narratives of aboriginal health inequity: Exploring Cree elder resilience and well-being in contexts of historical trauma. *Qualitative Health Research, 26*(14), 1911–1927. https://doi.org/10.1177/1049732315609569

Hayes, K., Blashki, G., Wiseman, J., Burke, S., & Reifels, L. (2018). Climate change and mental health: Risks, impacts and priority actions. *International Journal of Mental Health Systems, 12*, 28.

Hendriks, T., Schotanus-Dijkstra, M., Hassankhan, A., Graafsma, T., Bohlmeijer, E., & de Jong, J. (2018). The efficacy of positive psychology interventions from non-Western countries: A systematic review and meta-analysis. *International Journal of Wellbeing, 8*(1), 71–98. https://doi.org/10.5502/ijw.v8i1.711

Hendriks, T., Schotanus-Dijkstra, M., Hassankhan, A., Jong, J., & Bohlmeijer, E. (2020). The efficacy of multi-component positive psychology interventions: A systematic review and meta-analysis of randomized controlled trials. *Journal of Happiness Studies, 21*(1), 357–390.

Hendriks, T., Warren, M. A., Schotanus-Dijkstra, M., Hassankhan, A., Graafsma, T., Bolmeijer, E., & de Jong, J. (2019). How WEIRD are positive psychology interventions? A bibliometric analysis of randomized controlled trials on the science of well-being. *The Journal of Positive Psychology, 14*(4), 489–501.

Higgins, E. T. (1997). Beyond pleasure and pain. *American Psychologist, 52*(12), 1280–1300. https://doi.org/10.1037/0003-066X.52.12.1280

Hodge, A. S., Hook, J. N., Van Tongeren, D. R., Davis, D. E., & McElroy-Heltzel, S. E. (2020). Political humility: Engaging others with different political perspectives. *The Journal of Positive Psychology, 16*, 1–10. https://doi.org/10.1080/17439760.2020.1752784

Hsiang, S., Burke, M., & Miguel, E. (2013). Quantifying the influence of climate on human conflict. *American Association for the Advancement of Science, 341*, 1235367.

Janoff-Bulman, R., & Carnes, N. C. (2013). Surveying the moral landscape: Moral motives and group-based moralities. *Personality and Social Psychology Review, 17*(3), 219–236. https://doi.org/10.1177/1088868313480274

Kemp, A. H., Arias, J. A., & Fisher, Z. (2017). Social ties, health and wellbeing: A literature review and model. In A. Ibáñez, L. Sedeño, & A. García (Eds.), *Neuroscience and Social Science* (pp. 397–427). Springer. https://doi.org/10.1007/978-3-319-68421-5_17

Kern, M. L., Williams, P., Spong, C., Colla, R., Sharma, K., Downie, A., et al. (2020). Systems informed positive psychology. *The Journal of Positive Psychology, 15*(6), 705–715. https://doi.org/10.1080/17439760.2019.1639799

Kim, H., Doiron, K. M., Warren, M. A., & Donaldson, S. I. (2018). The international landscape of positive psychology research: A systematic review. *International Journal of Wellbeing, 8*(1), 50–70. https://doi.org/10.5502/ijw.v8i1.651

Koohsari, M. J., Sugiyama, T., Sahlqvist, S., Mavoa, S., Hadgraft, N., & Owen, N. (2015). Neighborhood environmental attributes and adults' sedentary behaviors: Review and research agenda. *Preventive Medicine, 77*, 141–149. https://doi.org/10.1016/j.ypmed.2015.05.027

Lai, C. K., Hoffman, K. M., & Nosek, B. A. (2013). Reducing implicit prejudice. *Social and Personality Psychology Compass, 7*(5), 315–330. https://doi.org/10.1111/spc3.12023

Lai, B. S., Kelley, M. L., Harrison, K. M., Thompson, J. E., & Self-Brown, S. (2015). Posttraumatic stress, anxiety, and depression symptoms among children after hurricane Katrina: A latent profile analysis. *Journal of Child and Family Studies, 24*(5), 1262–1270.

Lambert, L., Lomas, T., van de Weijer, M. P., Passmore, H. A., Joshanloo, M., Harter, J., et al. (2020). Towards a greater global understanding of wellbeing: A proposal for a more inclusive measure. *International Journal of Wellbeing, 10*(2), 1–18.

Lambert, L., Warren, M., Brulé, G., O'Brien, C., Murray, S., Mulay-Shah, A., et al. (2020). Perspectives: Using positive psychology and the United Nations' sustainable development goals to build a better world. *Middle East Journal of Positive Psychology, 6*, 1–28.

Lindahl, J. F., & Grace, D. (2015). The consequences of human actions on risks for infectious diseases: A review. *Infection Ecology & Epidemiology, 5*, 30048. https://doi.org/10.3402/iee.v5.30048

Liu, J., Potter, T., & Zahner, S. (2020). Policy brief on climate change and mental health/well-being. *Nursing Outlook, 68*(4), 517–522.

Lomas, T., & Ivtzan, I. (2016). Second wave positive psychology: Exploring the positive–negative dialectics of wellbeing. *Journal of Happiness Studies, 17*(4), 1753–1768.

Lomas, T., Waters, L., Williams, P., Oades, L. G., & Kern, M. L. (2020). Third wave positive psychology: Broadening towards complexity. *The Journal of Positive Psychology, 16*, 660. https://doi.org/10.1080/17439760.2020.1805501

Lucette, A., Ironson, G., Pargament, K. I., & Krause, N. (2016). Spirituality and religiousness are associated with fewer depressive symptoms in individuals with medical conditions. *Psychosomatics, 57*(5), 505–513. https://doi.org/10.1016/j.psym.2016.03.005

Mallett, R. K., & Monteith, M. J. (2019). *Confronting prejudice and discrimination: The science of changing minds and behaviors*. Academic Press. https://doi.org/10.1016/C2017-0-01959-6

Manning, C., & Clayton, S. (2018). Threats to mental health and wellbeing associated with climate change. In S. Clayton & C. Manning (Eds.), *Psychology and Climate Change: Human Perceptions, Impacts, and Responses* (pp. 217–244). Elsevier.

Marone, F. (2021). Hate in the time of coronavirus: Exploring the impact of the COVID-19 pandemic on violent extremism and terrorism in the West. *Security Journal, 35*, 205. https://doi.org/10.1057/s41284-020-00274-y

Mayer, C. H., & May, M. S. (2021). Women leaders transcending the demands of COVID-19: A positive psychology 2.0 perspective. *Frontiers in Psychology, 12*, 647658. https://doi.org/10.3389/fpsyg.2021.647658

Mead, J., Fisher, Z., & Kemp, A. (2021). Moving beyond disciplinary silos towards a transdisciplinary model of wellbeing: An invited review. *Frontiers in Psychology, 12*, 642093. https://doi.org/10.3389/fpsyg.2021.642093

Mead, J., Fisher, Z., Wilkie, L., Gibbs, K., Pridmore, J., Tree, J., et al. (2019). *Rethinking wellbeing: Toward a more ethical science of wellbeing that considers current and future generations*. Authorea. https://doi.org/10.22541/au.156649190.08734276

Meyer, S., & Warren, M. A. (2021). Exploring the role of character strengths in the endorsement of gender equality and pro-environmental action in the UAE. *Middle East Journal of Positive Psychology, 7*. https://middleeastjournalofpositivepsychology.org/index.php/mejpp/article/view/119

Oyserman, D. (2015). Psychology of values. In J. D. Wright (Ed.), *International Encyclopedia of the Social & Behavioral Sciences* (pp. 36–40). Elsevier. https://doi.org/10.1016/B978-0-08-097086-8.24030-0

Paluck, E. L., Porat, R., Clark, C. S., & Green, D. P. (2020). Prejudice reduction: Progress and challenges. *Annual Review of Psychology, 72*, 533–560. https://doi.org/10.1146/annurev-psych-071620-030619

Paremoer, L., Nandi, S., Serag, H., & Baum, F. (2021). Covid-19 pandemic and the social determinants of health. *British Medical Journal, 372*, n129. https://doi.org/10.1136/bmj.n129

Parks, A. C., & Biswas-Diener, R. (2013). Positive interventions: Past, present, and future. In T. B. Kashdan & J. Ciarrochi (Eds.), *Mindfulness, Acceptance, and Positive Psychology: The Seven Foundations of Well-Being* (pp. 140–165). New Harbinger.

Pasha-Zaidi, N., Warren, M. A., El Ashmawi, Y., & Kowai-Bell, N. (2021). Promoting allyship among south Asian and Arab Muslims toward black and Latino/a Muslims in American Islamic centers. In N. Pasha-Zaidi (Ed.), *Toward a Positive Psychology of Islam and Muslim: Spirituality, Struggle, and Social Justice* (pp. 307–331). Springer.

Passmore, H.-A., & Holder, M. D. (2017). Noticing nature: Individual and social benefits of a two-week intervention. *The Journal of Positive Psychology, 12*(6), 537–546. https://doi.org/10.1080/17439760.2016.1221126

Perry, B. L., Aronson, B., & Pescosolido, B. A. (2021). Pandemic precarity: COVID-19 is exposing and exacerbating inequalities in the American heartland. *Proceedings of the National Academy of Sciences, 118*(8), e2020685118. https://doi.org/10.1073/pnas.2020685118

Peterson, C., & Seligman, M. E. P. (2004). *Character strengths and virtues: A handbook and classification*. Oxford University Press.

Plant, E. A., & Butz, D. A. (2006). The causes and consequences of an avoidance-focus for interracial interactions. *Personality and Social Psychology Bulletin, 32*(6), 833–846. https://doi.org/10.1177/0146167206287182

Plant, E. A., Devine, P. G., & Peruche, M. B. (2010). Routes to positive interracial interactions: Approaching egalitarianism or avoiding prejudice. *Personality and Social Psychology Bulletin, 36*(9), 1135–1147. https://doi.org/10.1177/0146167210378018

Psaila, C. (2019). Spirituality: The cornerstone of wellbeing. In S. Vella, R. Falzon, & A. Azzopardi (Eds.), *Perspectives on Wellbeing* (pp. 35–45). Brill Sense.

Putra, I., Astell-Burt, T., Cliff, D. P., Vella, S. A., John, E. E., & Feng, X. (2020). The relationship between green space and prosocial behaviour among children and adolescents: A systematic review. *Frontiers in Psychology, 11*, 859.

Rao, M. A., & Donaldson, S. I. (2015). Expanding opportunities for diversity in positive psychology: An examination of gender, race, and ethnicity. *Canadian Psychology, 56*(3), 271. https://doi.org/10.1037/cap0000036

Raymond, I. J., & Raymond, C. M. (2019). Positive psychology perspectives on social values and their application to intentionally delivered sustainability interventions. *Sustainability Science, 14*(5), 1381–1393.

Sanchez, C., & Dunning, D. (2021). The anti-scientists bias: The role of feelings about scientists in COVID-19 attitudes and behaviors. *Journal of Applied Social Psychology, 51*(4), 461–473. https://doi.org/10.1111/jasp.12748

Schleussner, C.-F., Donges, J. F., Donner, R. V., & Schellnhuber, H. J. (2016). Armed-conflict risks enhanced by climate-related disasters in ethnically fractionalized countries. *Proceedings of the National Academy of Sciences of the*

United States of America, 113, 9216–9221. https://doi.org/10.1073/pnas.1601611113

Schutte, N. S., & Malouff, J. M. (2019). The impact of signature character strengths interventions: A meta-analysis. *Journal of Happiness Studies, 20*(4), 1179–1196. https://doi.org/10.1007/s10902-018-9990-2

Seligman, M. E. P., Steen, T., Park, N., & Peterson, C. (2005). Positive psychology progress: Empirical validation of interventions. *American Psychologist, 60*(5), 410–421.

Slobodin, O., & Cohen, O. (2020). A culturally-competent approach to emergency management: What lessons can we learn from the COVID-19? *Psychological Trauma: Theory, Research, Practice, and Policy, 12*(5), 470–473. https://doi.org/10.1037/tra0000790

Tajfel, H. (1981). *Human groups and social categories*. Cambridge University Press.

Tal, A., & Kerret, D. (2020). Positive psychology as a strategy for promoting sustainable population policies. *Heliyon, 6*(4), e03696. https://doi.org/10.1016/j.heliyon.2020

Taylor, M. S., Wheeler, B. W., White, M. P., Economou, T., & Osborne, N. J. (2015). Research note: Urban street tree density and antidepressant prescription rates - A cross-sectional study in London, UK. *Landscape and Urban Planning, 136*, 174–179.

Towers, S., Chen, S., Malik, A., & Ebert, D. (2018). Factors influencing temporal patterns in crime in a large American city: A predictive analytics perspective. *PLoS One, 13*(10), e0205151. https://doi.org/10.1371/journal.pone.0205151

Trawalter, S., & Richeson, J. A. (2006). Regulatory focus and executive function after interracial interactions. *Journal of Experimental Social Psychology, 42*(3), 406–412. https://doi.org/10.1016/j.jesp.2005.05.008

Ward Thompson, C., Aspinall, P., Roe, J., Robertson, L., & Miller, D. (2016). Mitigating stress and supporting health in deprived urban communities: The importance of green space and the social environment. *International Journal of Environmental Research and Public Health, 13*(4), 440. https://doi.org/10.3390/ijerph13040440

Warren, M. A., & Bordoloi, S. D. (2020). When COVID-19 exacerbates inequities: The path forward for generating wellbeing. *International Journal of Wellbeing, 10*(3), 1–6. https://doi.org/10.5502/ijw.v10i3.1357

Warren, M. A., Sekhon,T., & Waldrop, R. J. (2022). Highlighting strengths in response to discrimination: Developing and testing an allyship positive psychology intervention. *International Journal of Wellbeing, 12*(1), 21–41. https://doi.org/10.5502/ijw.v12i1.1751

Waters, L., Cameron, K., Nelson-Coffey, K., Crone, D. L., Kern, M. L., Lomas, T., et al. (2021). Collective wellbeing and posttraumatic growth during COVID-19: How positive psychology can help families, schools, workplaces

and marginalized communities. *The Journal of Positive Psychology*. https://doi.org/10.1080/17439760.2021.1940251
Weinstein, N., Balmford, A., DeHaan, C. R., Gladwell, V., Bradbury, R. B., & Amano, T. (2015). Seeing community for the trees: The links among contact with natural environments, community cohesion, and crime. *Bioscience, 65*, 1141–1153.
White, M. P., Alcock, I., Grellier, J., Wheeler, B. W., Hartig, T., Warber, S. L., et al. (2019). Spending at least 120 minutes a week in nature is associated with good health and wellbeing. *Scientific Reports, 9*(1), 7730. https://doi.org/10.1038/s41598-019-44097-3
Wong, P. T. P. (2011). Positive psychology 2.0: Towards a balanced interactive model of the good life. *Canadian Psychology/Psychologie Canadienne, 52*, 69–81. https://doi.org/10.1037/a0022511
Yaden, D. B., Claydon, J., Bathgate, M., Platt, B., & Santos, L. R. (2021). Teaching well-being at scale: An intervention study. *PLoS One, 16*(4), e0249193. https://doi.org/10.1371/journal.pone.0249193
Zabel, K. L., Olson, M. A., Johnson, C. S., & Phillips, J. E. (2015). What we talk about matters: Content moderates cognitive depletion in interracial interactions. *The Journal of Social Psychology, 155*(6), 545–552. https://doi.org/10.1080/00224545.2015.1032197
Zelenski, J. M., Dopko, R. L., & Capaldi, C. A. (2015). Cooperation is in our nature: Nature exposure may promote cooperative and environmentally sustainable behavior. *Journal of Environmental Psychology, 42*, 24–31.

CHAPTER 4

Towards a Culture of Care for Societal Wellbeing: A Perspective from the Healthcare Sector

Katie Gibbs, Zoe Fisher, and Andrew H. Kemp

Abstract This chapter describes some of the challenges facing the healthcare sector which have led us to reflect on how we might redesign and transform healthcare services. We have developed a theoretical framework of wellbeing on which we have developed previously unimagined interventions for delivery in a community brain injury service in South Wales in the United Kingdom. This framework is inspired by social ecological

K. Gibbs
School of Psychology, Faculty of Medicine, Health and Life Science, Swansea University, Swansea, Wales, UK

Z. Fisher
Regional Neuropsychology and Community Brain Injury Service, Swansea Bay University Health Board, Morriston Hospital, Swansea, Wales, UK

Health and Wellbeing Academy, Faculty of Medicine, Health and Life Science, Swansea University, Swansea, Wales, UK

A. H. Kemp, D. J. Edwards (eds.), *Broadening the Scope of Wellbeing Science*, https://doi.org/10.1007/978-3-031-18329-4_4

theory and systems-informed thinking in which the individual is positioned within increasing phenomenological scales, extending to the ecosystem and the life course, highlighting systemic influences on wellbeing that also change over time. We present a concrete example from our own work relating to how we have begun to promote a culture of care that extends beyond the healthcare sector into the community and natural environment. Key implications of our work for building societal wellbeing include the need to: (1) promote wellbeing which is not the same as reducing 'ill-being', (2) support the person to adapt and function well within the context of the wider systems in which the person is embedded, (3) foster cultures of care through strategic partnerships across organisations, and (4) commit to the science and art of system transformation with an eye on co-production of knowledge and co-creation of opportunity. A focus on wellbeing lays the foundation for improving 'whole health' at scale, involving a deeper appreciation for individual, collective and planetary wellbeing.

Keywords GENIAL • Social ecology • Systems-informed • Individual wellbeing • Collective wellbeing • Planetary wellbeing

The increasing burden of physical ill-health and mental ill-being is placing unprecedented pressure on healthcare systems and practices (Global Burden of Disease Collaborators, 2016; Whiteford et al., 2013). Here we describe some of the challenges facing the healthcare sector that had led us to develop a theoretical framework of 'wellbeing' (Kemp et al., 2017; Kemp & Fisher, 2022; Mead et al., 2019, 2021), which we have defined as the sense of connectedness to ourselves (the individual domain), others (the community domain), and nature (the environment domain), and a basic psychological need that is supported by vagal function, a

---

A. H. Kemp (✉)
School of Psychology, Faculty of Medicine, Health and Life Science, Swansea University, Swansea, Wales, UK

Regional Neuropsychology and Community Brain Injury Service, Swansea Bay University Health Board, Morriston Hospital, Swansea, Wales, UK
e-mail: a.h.kemp@swansea.ac.uk

psychophysiological resource for connection (Kemp & Fisher, 2022; Mead et al., 2022; Wilkie et al., 2022). A focus on wellbeing is relevant to the healthcare sector because it has been shown to causally affect health and longevity, even after controlling for health and socioeconomic status at baseline (e.g. Diener & Chan, 2011). Our work has been inspired by developments in the field (Lomas et al., 2020; Mead et al., 2019; Wong, 2019) as well as social ecological theory, which places the person in the context of the wider systems in which they are embedded (Bronfenbrenner, 1977; Fisher et al., 2020; Lomas, 2015) and systems-informed thinking, which draws on concepts from the systems sciences in order to optimise human social systems (Kern et al., 2020; Leischow et al., 2008; Willis et al., 2014). We elaborate on these developments below and illustrate how our framework has led to significant clinical service redesign in a community neurorehabilitation service in South Wales in the United Kingdom, which has enabled the realisation of previously unimagined opportunities. We end our chapter by providing a concrete example of one of these initiatives.

Reflecting on the Challenges Facing Healthcare

Chronic conditions have now superseded acute conditions as leading burdens of morbidity, mortality and healthcare expenditures (Bolnick et al., 2020; GBD Collaborators, 2020). Although we are living longer, we are living with more disease, infirmity and multi-morbidity. While medical research has led to an increase in life expectancy, a focus on quantity—rather than quality of life—has led to an increase in age-related disease and societal burden (Brown, 2015; Garmany, Yamada, & Terzic, 2021). In the UK, it is estimated that by 2035, 17% of the population will have four or more chronic conditions; double that of current prevalence levels (9.8%) (Kingston et al., 2018). This poses a real challenge for the single disease-oriented model which is woven into the fabric of our healthcare institutions (Rocca & Anjum, 2020; Wang et al., 2019). Unfortunately, our models of healthcare have not adapted to reflect the epidemiological transition from acute to chronic conditions (GBD Collaborators, 2020; Kemp et al., 2022; Kruk et al., 2018). Inherent in our models of care is the erroneous assumption that the reduction of ill-health (ill-being) equates to health (wellbeing), meaning that critical opportunities to improve health and wellbeing are being missed (Fisher et al., 2020; Rocca & Anjum, 2020). This issue is compounded by the principle of dualism—that physical and mental health are separable—a principle that is enshrined in the infrastructure of our healthcare institutions (Switankowsky, 2000; Latoo et al., 2021), resulting in unhelpful silos between disciplines, constraining more holistic and evidence-based approaches (Burgmer & Forstmann, 2018). We suggest that this does not need to be the case, but also argue that system transformation is needed to overcome entrenched barriers to change.

Although multi-morbidity has been recognised as an urgent priority for healthcare research, people with multi-morbidity are often excluded from clinical trials due to increased heterogeneity (du Vaure et al., 2016). People living with multi-morbidity report experiencing fragmented and disjointed care as well as poorer outcomes despite increased total costs, hospital costs, care transition costs, primary care use, dental care use, emergency department use, and hospitalisations (Whitty, 2017; Whitty et al., 2020). Research has also highlighted that service utilisation and costs of providing those services are higher in patients with multi-morbidity, when accompanied by depression and distress (Soley-Bori et al., 2021; Suls et al., 2019), highlighting a need for greater attention to biopsychosocial issues in a systems-informed way.

Our own focus on chronic conditions has concentrated on acquired brain injury (ABI) because it is the area in which we work clinically. An estimated 1.3 million people live with the effects of brain injury at a cost to the UK economy of GBP 15 billion per annum; equivalent to 10% of the annual NHS budget (Barber et al., 2018). ABI can lead to a range of physical, cognitive, psychological and social difficulties. Critically, ABI is associated with a host of physical multi-morbidities as well as significant long-term emotional distress, depressive disorders (Glenn et al., 2001; Jorge et al., 2004; Osborn et al., 2014) and a fourfold increased risk of suicide (Mackelprang et al., 2014; Teasdale & Engberg, 2001). Despite this, people living with ABI typically have little access to psychosocial interventions to address ongoing holistic needs, with research suggesting that up to three-quarters of people living with ABI feel that their psychological needs are not being met by traditional models of care (McKevitt et al., 2011). Accordingly, the focus of our work has been to develop more holistic models of care to better meet the needs of people living with ABI (see Fisher et al., 2020) and now reflect on the salient points applicable to the care of people living with chronic conditions and multi-morbidity generally.

Building More Holistic Models of Healthcare

Our work has involved developing our own theoretical model of wellbeing, around which we have developed new interventions for promoting wellbeing in the healthcare sector (see Kemp & Fisher, 2022 for review; see also Fig. 4.1). Our theoretical work was largely inspired by the pragmatic need to understand what wellbeing actually was in order to design a clinical service that could facilitate it, rather than simply aiming to reduce deficits and distress. To that end, our 'GENIAL' model was derived by imposing an interpretative framework on the published literature working towards a transdisciplinary understanding of wellbeing (Fisher et al., 2020; Kemp et al., 2017; Kemp & Fisher, 2022; Mead et al., 2021) explicitly linked to a broader context including sociocontextual factors. The

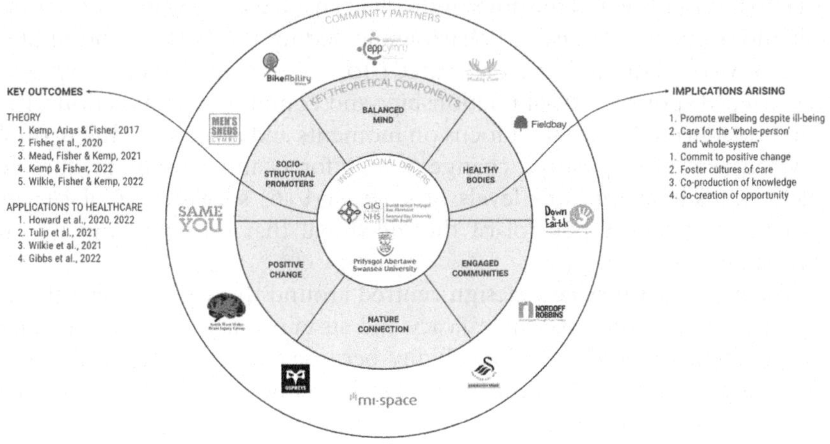

Fig. 4.1 A 'GENIAL' Approach to Wellbeing. Our approach to wellbeing focuses on core pillars that we have identified from the peer-reviewed literature and on which we have built our theoretical framework ('GENIAL'). Our approach is driven by a collaborative partnership between the university and healthcare sector, supported by further strategic partnerships with community-based organisations. These partnerships have led to key outcomes in the peer-reviewed literature reporting development of theory and associated applications from which key implications for the transformation of the healthcare sector have arisen

GENIAL acronym refers to relationships between *G*enomics-*E*nvironment-vagus *N*erve-social *I*nteraction-*A*llostatic regulation-*L*ongevity which over the life course support or impede wellbeing. Our framework 'brings together' other influential models across different disciplines, integrating heterogeneous ideas into a coherent whole. This framework provides the theoretical glue to connect disparate ideas, illustrating how these ideas reciprocally interact to realise wellbeing at multiple levels of scale, spanning the individual, community and environment.

To promote wellbeing, we have identified core pillars around which we have developed several new interventions (see Kemp & Fisher, 2022 for a more detailed overview). With regard to the long-term management of chronic conditions, we suggest that designing services around these pillars may present opportunities to increase the efficacy and sustainability of models of care and improve the health and wellbeing at scale. These pillars include the promotion of (1) emotional balance through, for example, acceptance of difficult emotions, individual character strengths, and meaning and purpose, (2) health behaviours including the prescription of lifestyle advice relating (but not limited) to, for example, diet, sleep and exercise, (3) positive social relationships and community connections, (4) nature connectedness with a focus on moments and minutes in the natural environment, (5) positive change to reinforce and sustain favourable changes made at multiple levels of scale, and (6) sociostructural factors that lie beyond the control of the individual that positively impact on wellbeing.

We argue that service redesign centred around these key pillars affords opportunities to increase the efficacy and sustainability of models of care for people living with multi-morbidity because they promote health and wellbeing, rather than concentrating on the reduction of deficits and distress. A focus on emotional balance is supported by developments in psychological science that is now embracing the bright and dark sides of life in order to achieve self-transcendence, broadly defined (see chapter 2 in this book). A complementary focus on positive health behaviours seeks to place 'body' and 'mind' on equal footing in regard to the promotion of wellbeing (Gibbs et al., 2022; Wilkie et al., 2021 for specific examples). Recognising a critical role for social ties and the natural environment in the promotion of wellbeing (Haslam et al., 2017; Kemp et al., 2017; Martin et al., 2020; Kemp & Fisher, 2022; Wilkie et al., 2021) also lays foundations for collective and planetary wellbeing (Antó et al., 2021; JYU.Wisdom Community, 2021; Waters et al., 2021). Inclusion of behaviour change recognises the importance of sustaining positive change at multiple levels of scale to overcome the so-called intention-behaviour gap (Cappellen et al., 2017; Fabian & Pykett, 2021). The final pillar of 'sociostructural promoters' of wellbeing recognises that there are often constraints to improving wellbeing beyond the control of the individual (e.g., socioeconomic inequalities) and clinical services (resource constraints), highlighting the need to identify wider contextual facilitators or promoters of wellbeing to support sustained change (e.g., strategic public health strategies to reduce health inequalities).

Critically, the development of more effective solutions to the global burden of disease and improving the health and wellbeing of the population cannot be seen as the sole remit of health services. This needs to be a societal endeavour that is addressed at multiple levels of scale. To that end, we now reflect on how systems-informed thinking might lead to more effective and sustainable solutions.

Systems-Informed Promotion of Wellbeing

Wellbeing has been described as a 'wicked problem' (Bache et al., 2016) as it has been difficult to define and has avoided straightforward solutions. The complexity of wellbeing is illustrated nicely through Bronfenbrenner's social ecological model and its adaptations (Bronfenbrenner, 1977; Lomas, 2015) which have inspired our own work in this area, focused on the healthcare sector. According to social ecological frameworks, the individual is embedded within multiple microsystems relating to the immediate social settings of the individual including the family, school and workplace. The interaction between these social settings is itself a system (the mesosystem), which is impacted on by wider social structures (the exosystem), all of which are influenced by systemic economic, educational and political systems (the macrosystem), and the biosphere on which all other levels depend (the ecosystem). This social ecological framework highlights the considerable complexity and interdependent influences, promoters and constraints on the capacity for wellbeing to arise in different contexts over time.

Systems-informed thinking lends itself to addressing such problems because it takes into account the complexity and the interdependence of related and overlapping systems (Kreuter et al., 2004). This approach encourages a focus on dynamic relationships, complex interactions and partnership working to facilitating self-organisational capacity through adaptation and emergent change (Holmes et al., 2017; Kern et al., 2020; Leischow et al., 2008) laying foundations for system transformation. We have sought to capture this complexity in our own multi-levelled framework, drawing on social ecological theory, placing the individual within increasingly higher levels of scale relating to the systems in which people are embedded. While our framework is a simplified example of a social ecological model spanning the individual, community and environment, it has still been useful in terms of helping us develop our interventions (see Kemp & Fisher, 2022 for a comprehensive review).

Systems theory has encompassed a focus on 'hard' and 'soft' systems (Kern et al., 2020), in which the 'hard' systems tradition has focused on developing computer models to better understand and quantify the role of feedback and dynamic relationships within the system. In contrast, 'soft' systems thinking is a more phenomenological and interpretative approach involving open dialogue around complex problems, drawing on multiple perspectives and stakeholders (Kern et al., 2020). This 'soft' systems approach has evolved from a need to focus on human social systems, which is more difficult to measure using a quantitative approach. Elements within and across systems interact with important implications for health and wellbeing, both in terms of the problem (e.g., the disease burden attributable to chronic conditions on individuals, families and societies) and the solution (e.g. strategies for managing that burden by improving the capacity of individuals to improve their wellbeing while harnessing societal and community assets through collaboration and partnership working).

A systems approach to healthcare is a topic of great research interest and has been described as "*a way of addressing health delivery challenges that recognises the multiplicity of elements interacting to impact an outcome of interest and implements processes or tools in a holistic way*" (Komashie et al., 2021). While application of systems-thinking is often theoretical in nature, our work joins other's efforts (e.g. Holmes et al., 2017) to move this discussion to a more practical level, demonstrating a 'shared purpose' seeking to promote wellbeing at scale with an eye on creating previously unimagined positive futures (Kern et al., 2020). Our own service transformation has focused on driving forward previously unexplored initiatives (see Kemp & Fisher, 2022 for review) supported by strategic partnerships with community-based organisations leading to co-creation of opportunities for promoting health and wellbeing which overcome resource constraints and siloed working. A strategic partnership between the local area health board and university has allowed us to embed research activity into the clinical setting (Boaz et al., 2015; Friedman et al., 2015; Grumbach et al., 2014; Ozdemir et al., 2015), facilitating knowledge to action, a challenge that has plagued clinical and health service research (e.g. Grimshaw et al., 2012; Heneghan et al., 2017). We have observed how these initiatives have provided service users with avenues for resilience and sustainable wellbeing long after discharge from the health service. We now provide a concrete example of a co-produced surf therapy intervention between a health service, university and a community organisation.

A Concrete Example of System Transformation

We focus here on one concrete example involving surf therapy for individuals living with ABI, arising from a strategic partnership across university, health service and community providers (Gibbs et al., 2022). It is important to point out here not all people living with ABI want to engage with surf therapy, but—guided by a patient-centred approach to care (Institute of Medicine (US) Committee on Quality of Health Care in America, 2001)—for those that do or those at least willing and physically able to give it a try, we ensure that such an opportunity is able to be realised through strategic partnerships across organisations, supporting the service user to realise new opportunities that they have previously believed that they were no longer able to enjoy. A recently published qualitative evaluation of this work (Gibbs et al., 2022) highlighted seven overarching themes, including: (1) Connection to Nature, (2) Facilitating Trust and Safety, (3) Managing and Accepting Difficult Emotions, (4) Facilitating Positive Emotion, Meaning and Purpose, (5) Building Community through Social Connection, and (6) Positive Change. Barriers and Opportunities (7) were also identified, providing insights into how clinical services may be improved.

Surf therapy provided the context for the experience of multiple determinants of wellbeing in line with our GENIAL theoretical framework (Fisher et al., 2020; Kemp et al., 2017; Kemp & Fisher, 2022; Mead et al., 2021). For instance, at the level of the individual, surf therapy created opportunities for managing difficult emotions, drawing on their character strengths and experiencing positive emotions while engaging in positive health behaviours. Participants learnt that unpleasant events and feelings arising from unpredictable and uncontrollable conditions are not always able to be controlled, but that these same conditions may also promote feelings of revitalisation, refreshment and vigour, facilitating emotional balance during outdoor physical activity. At the level of community, surf therapy provided opportunities to overcome loneliness and foster meaningful social ties with others from a diverse array of backgrounds. By learning new skills and sharing experiences with similar others, participants experienced a sense of belonging and social identity, further promoting positive psychological and physical health outcomes. At the level of the environment, the coastal setting provided the context for connecting to the natural environment, contributing to a sense of relaxation, revitalisation and refreshment, and an appreciation of beauty, with which individual and collective wellbeing was realised. Positive behaviour change was promoted within and beyond the experience of surfing, reinforcing new positive habits that serve to shift the mindset of participants from ailments to capabilities. This work is now feeding into discussions with charities and government as to how we might better promote societal wellbeing at scale, drawing on our bottom-up approach described here, but also feeding into top-down opportunities through, for example, influencing wellbeing public policy (Fabian & Pykett, 2021).

This project provides an exemplar of what can be achieved by fostering and nurturing relationships within and between organisations in order to create, deliver and evaluate evidence-based models of care which seamlessly utilise the skills and resources of all parties to enhance individual and community wellbeing. As shown in Fig. 4.1, surf therapy is one example of many initiatives that have been co-created to provide service users with a range of diverse opportunities for community integration to suit their interests and needs. Partnership working allows for disparate assets to be brought together into a coherent whole to facilitate integrated working towards individual, societal, and planetary wellbeing. Importantly, establishing effective partnerships between healthcare services, local communities, academic and third-sector organisations presents opportunities to facilitate the delivery of various evidence-based wellbeing interventions that would not otherwise be available or accessible to individuals.

Discussion and Conclusion

This chapter has explored some of the challenges facing the healthcare sector and opportunities to widen the lens for tackling chronic conditions and building societal wellbeing. Our framework has several important implications for the healthcare sector, including the need to: (1) promote wellbeing despite considerable 'ill-being', (2) focus on the 'whole person' rather than the diseased body, (3) commit to positive change to overcome the disconnect between what people know and what they actually do, (4) foster cultures of care through strategic partnerships with community-based organisations, and (5) contribute to the science and art of system change, facilitating the co-production of knowledge and co-creation of opportunity.

References

Antó, M., Martí, L., Casals, J., Bou-Habib, P., Casal, P., Fleurbaey, M., Frumkin, H., Jiménez-Morales, M., Jordana, J., Lancelotti, C., Llavador, C., Mélon, L., Solé, R., Subirada, F., & Williams, A. (2021). The planetary wellbeing initiative: Pursuing the sustainable development goals in higher education. *Sustainability, 13*(6), 3372. https://doi.org/10.3390/su13063372

Bache, I., Reardon, L., & Anand, P. (2016). Wellbeing as a wicked problem: Navigating the arguments for the role of government. *Journal of Happiness Studies, 17*(3), 893–912. https://doi.org/10.1007/s10902-015-9623-y

Barber, S., Bate, A., Adcock, A., Harker, R., Roberts, N., & Woodhouse, J. (2018). *House of Commons Library, debate pack: Acquired brain injury*. House of Commons.

Boaz, A., Hanney, S., Jones, T., & Soper, B. (2015). Does the engagement of clinicians and organisations in research improve healthcare performance: A three-stage review. *BMJ Open, 5*(12), e009415. https://doi.org/10.1136/bmjopen-2015-009415

Bolnick, H. J., Bui, A. L., Bulchis, A., Chen, C., Chapin, A., Lomsadze, L., Mokdad, A. H., Millard, F., & Dieleman, J. L. (2020). Health-care spending attributable to modifiable risk factors in the USA: An economic attribution analysis. *The Lancet Public Health, 5*(10), e525–e535. https://doi.org/10.1016/s2468-2667(20)30203-6

Bronfenbrenner, U. (1977). Toward an experimental ecology of human development. *American Psychologist, 32*(7), 513–531. https://doi.org/10.1037/0003-066x.32.7.513

Brown, G. C. (2015). Living too long. *EMBO Reports, 16*(2), 137–141. https://doi.org/10.15252/embr.201439518

Burgmer, P., & Forstmann, M. (2018). Mind-body dualism and health revisited: How belief in dualism shapes health behavior. *Social Psychology, 49*(4), 219–230. https://doi.org/10.1027/1864-9335/a000344

Cappellen, P. V., Rice, E. L., Catalino, L. I., & Fredrickson, B. L. (2017). Positive affective processes underlie positive health behaviour change. *Psychology & Health, 33*(1), 1–21. https://doi.org/10.1080/08870446.2017.1320798

GBD Collaborators. (2020). Global burden of 369 diseases and injuries in 204 countries and territories, 1990–2019: A systematic analysis for the Global Burden of Disease Study 2019. *The Lancet, 396*(10258), 1204–1222. https://doi.org/10.1016/s0140-6736(20)30925-9

Diener, E., & Chan, M. Y. (2011). Happy people live longer: Subjective well-being contributes to health and longevity. *Applied Psychology, 3*(1), 1–43. https://doi.org/10.1111/j.1758-0854.2010.01045.x

Fabian, M., and Pykett, J. (2021). Be Happy: Navigating Normative Issues in Behavioral and Well-Being Public Policy. *Perspectives on Psychological Science, 17*, 174569162098439. https://doi.org/10.1177/1745691620984395.

Fisher, Z., Galloghly, E., Boglo, E., Gracey, F., & Kemp, A. (2020). Emotion, wellbeing and the neurological disorders. In *Reference Module in Neuroscience and Biobehavioral Psychology*. Elsevier. https://doi.org/10.1016/b978-0-12-819641-0.00013-x.

Friedman, C., Rubin, J., Brown, J., Buntin, M., Corn, M., Etheredge, L., Gunter, C., Musen, M., Platt, R., Stead, W., Sullivan, K., & Houweling, D. V. (2015). Toward a science of learning systems: A research agenda for the high-functioning Learning Health System. *Journal of the American Medical Informatics Association, 22*(1), 43–50. https://doi.org/10.1136/amiajnl-2014-002977

Garmany, A., Yamada, S., & Terzic, A. (2021). Longevity leap: Mind the health span gap. *NPJ Regenerative Medicine, 6*(1), 57. https://doi.org/10.1038/s41536-021-00169-5

Gibbs, K., Wilkie, L., Jarman, J., Barker-Smith, A., Kemp, A. H., & Fisher, Z. (2022). Riding the wave into wellbeing: A qualitative evaluation of surf therapy for individuals living with acquired brain injury. *PLoS One, 17*(4), e0266388. https://doi.org/10.1371/journal.pone.0266388

Glenn, M., O'Neil-Pirozzi, T., Goldstein, R., Burke, D., & Jacob, L. (2001). Depression amongst outpatients with traumatic brain injury. *Brain Injury, 15*(9), 811–818. https://doi.org/10.1080/02699050120330

Global Burden of Disease Collaborators. (2016). Global, regional, and national life expectancy, all-cause mortality, and cause-specific mortality for 249 causes of death, 1980–2015: A systematic analysis for the Global Burden of Disease Study 2015. *The Lancet, 388*, 1459. https://doi.org/10.1016/s0140-6736(16)31012-1

Grimshaw, J. M., Eccles, M. P., Lavis, J. N., Hill, S. J., & Squires, J. E. (2012). Knowledge translation of research findings. *Implementation Science, 7*(1), 50. https://doi.org/10.1186/1748-5908-7-50

Grumbach, K., Lucey, C. R., & Johnston, S. C. (2014). Transforming from centers of learning to learning health systems: The challenge for academic health centers. *Journal of the American Medical Association, 311*(11), 1109–1110. https://doi.org/10.1001/jama.2014.705

Haslam, C., Jetten, J., Cruwys, T., Dingle, G., & Haslam, A. (2017). *The new psychology of health: Unlocking the social cure*. Routledge. https://www.routledge.com/The-New-Psychology-of-Health-Unlocking-the-Social-Cure/Haslam-Jetten-Cruwys-Dingle-Haslam/p/book/9781138123885

Heneghan, C., Goldacre, B., & Mahtani, K. R. (2017). Why clinical trial outcomes fail to translate into benefits for patients. *Trials, 18*(1), 122. https://doi.org/10.1186/s13063-017-1870-2

Holmes, B. J., Best, A., Davies, H., Hunter, D., Kelly, M. P., Marshall, M., & Rycroft-Malone, J. (2017). Mobilising knowledge in complex health systems: A call to action. *Evidence and Policy, 13*(3), 539–560. https://doi.org/10.1332/174426416x14712553750311

Institute of Medicine (US) Committee on Quality of Health Care in America. (2001). *Crossing the quality chasm: A new health system for the 21st century*. National Academies Press.

Jorge, R. E., Robinson, R. G., Moser, D., Tateno, A., Crespo-Facorro, B., & Arndt, S. (2004). Major depression following traumatic brain injury. *Archives of General Psychiatry, 61*(1), 42–50. https://doi.org/10.1001/archpsyc.61.1.42

JYU.Wisdom Community. (2021). Planetary well-being. *Humanities and Social Sciences Communications, 8*(1), 258. https://doi.org/10.1057/s41599-021-00899-3

Kemp, A. H., & Fisher, Z. (2022). Wellbeing, whole health and societal transformation: Theoretical insights and practical applications. *Global Advances in Health and Medicine, 11*, 216495612110730. https://doi.org/10.1177/21649561211073077

Kemp, A. H., Arias, J. A., & Fisher, Z. (2017). Social ties, health and wellbeing: A literature review and model. In *Neuroscience and Social Science, The Missing Link* (Vol. 59, pp. 397–427). Springer International Publishing. https://doi.org/10.1007/978-3-319-68421-5_17

Kemp, A. H., Tree, J., Gracey, F., & Fisher, Z. (2022). Editorial: Improving wellbeing in patients with chronic conditions: Theory, evidence, and opportunities. *Frontiers in Psychology, 13*, 868810. https://doi.org/10.3389/fpsyg.2022.868810

Kern, M. L., Williams, P., Spong, C., Colla, R., Sharma, K., Downie, A., Taylor, J. A., Sharp, S., Siokou, C., & Oades, L. G. (2020). Systems informed positive psychology. *The Journal of Positive Psychology, 15*(6), 705–715. https://doi.org/10.1080/17439760.2019.1639799

Kingston, A., Robinson, L., Booth, H., Knapp, M., Jagger, C., & project, M. (2018). Projections of multi-morbidity in the older population in England to 2035: Estimates from the Population Ageing and Care Simulation (PACSim) model. *Age and Ageing, 47*(3), 374–380. https://doi.org/10.1093/ageing/afx201

Komashie, A., Ward, J., Bashford, T., Dickerson, T., Kaya, G. K., Liu, Y., Kuhn, I., Günay, A., Kohler, K., Boddy, N., O'Kelly, E., Masters, J., Dean, J., Meads, C., & Clarkson, P. J. (2021). Systems approach to health service design, delivery and improvement: A systematic review and meta-analysis. *BMJ Open, 11*(1), e037667. https://doi.org/10.1136/bmjopen-2020-037667

Kreuter, M. W., Rosa, C. D., Howze, E. H., & Baldwin, G. T. (2004). Understanding wicked problems: A key to advancing environmental health promotion. *Health Education & Behavior, 31*(4), 441–454. https://doi.org/10.1177/1090198104265597

Kruk, M. E., Gage, A. D., Arsenault, C., Jordan, K., Leslie, H. H., Roder-DeWan, S., Adeyi, O., Barker, P., Daelmans, B., Doubova, S. V., English, M., Elorrio, E. G., Guanais, F., Gureje, O., Hirschhorn, L. R., Jiang, L., Kelley, E., Lemango, E. T., Liljestrand, J., et al. (2018). High-quality health systems in the Sustainable Development Goals era: Time for a revolution. *The Lancet. Global Health, 6*(11), e1196–e1252. https://doi.org/10.1016/s2214-109x(18)30386-3

Latoo, J., Mistry, M., Alabdulla, M., Wadoo, O., Jan, F., Munshi, T., Iqbal, Y., & Haddad, P. (2021). Mental health stigma: The role of dualism, uncertainty, causation and treatability. *General Psychiatry, 34*(4), e100498. https://doi.org/10.1136/gpsych-2021-100498

Leischow, S. J., Best, A., Trochim, W. M., Clark, P. I., Gallagher, R. S., Marcus, S. E., & Matthews, E. (2008). Systems thinking to improve the public's health. *American Journal of Preventive Medicine, 35*(2), S196–S203. https://doi.org/10.1016/j.amepre.2008.05.014

Lomas, T. (2015). Positive social psychology: A multilevel inquiry into sociocultural well-being initiatives. *Psychology Public Policy and Law, 21*(3), 338–347. https://doi.org/10.1037/law0000051

Lomas, T., Waters, L.,Oades, L. G., & Kern, M. L. (2020). Third wave positive psychology: Broadening towards complexity. *The Journal of Positive Psychology*, 1–15. https://doi.org/10.1080/17439760.2020.1805501.

Mackelprang, J. L., Bombardier, C. H., Fann, J. R., Temkin, N. R., Barber, J. K., & Dikmen, S. S. (2014). Rates and predictors of suicidal ideation during the first year after traumatic brain injury. *American Journal of Public Health, 104*(7), e100–e107. https://doi.org/10.2105/ajph.2013.301794

Martin, L., White, M. P., Hunt, A., Richardson, M., Pahl, S., & Burt, J. (2020). Nature contact, nature connectedness and associations with health, wellbeing and pro-environmental behaviours. *Journal of Environmental Psychology, 68*, 101389. https://doi.org/10.1016/j.jenvp.2020.101389

McKevitt, C., Fudge, N., Redfern, J., Sheldenkar, A., Crichton, S., Rudd, A. R., Forster, A., Young, J., Nazareth, I., Silver, L. E., Rothwell, P. M., & Wolfe, C. D. A. (2011). Self-reported long-term needs after stroke. *Stroke, 42*(5), 1398–1403. https://doi.org/10.1161/strokeaha.110.598839

Mead, J., Fisher, Z., & Kemp, A. H. (2021). Moving beyond disciplinary silos towards a transdisciplinary model of wellbeing: An invited review. *Frontiers in Psychology, 12*, 642093. https://doi.org/10.3389/fpsyg.2021.642093

Mead, J., Fisher, Z., Wilkie, L., Gibbs, K., Pridmore, J., Tree, J., & Kemp, A. (2019). Toward a more ethical science of wellbeing that considers current and future generations. *Authorea*. https://doi.org/10.22541/au.156649190.08734276

Mead, J., Playfoot, D., Fisher, Z., Tree, J., & Kemp, A. H. (2022). Pathways to post-traumatic growth and wellbeing during the COVID-19 pandemic: Findings from a UK-based sample. *International Journal of Existential Positive Psychology, 11*, 1–15. Available at: https://www.meaning.ca/ijepp-article/vol11-no1/pathways-to-post-traumatic-growth-and-wellbeing/

Osborn, A. J., Mathias, J. L., & Fairweather-Schmidt, A. K. (2014). Depression following adult, non-penetrating traumatic brain injury: A meta-analysis examining methodological variables and sample characteristics. *Neuroscience & Biobehavioral Reviews, 47*, 1–15. https://doi.org/10.1016/j.neubiorev.2014.07.007

Ozdemir, B. A., Karthikesalingam, A., Sinha, S., Poloniecki, J. D., Hinchliffe, R. J., Thompson, M. M., Gower, J. D., Boaz, A., & Holt, P. J. E. (2015). Research activity and the association with mortality. *PLoS One, 10*(2), e0118253. https://doi.org/10.1371/journal.pone.0118253

Rocca, E., & Anjum, R. L. (2020). Complexity, Reductionism and the Biomedical Model. In Anjum, R., Copeland, S., Rocca, E. (Eds.) *Rethinking Causality, Complexity and Evidence for the Unique Patient*. Springer, Cham. https://doi.org/10.1007/978-3-030-41239-5_5

Soley-Bori, M., Ashworth, M., Bisquera, A., Dodhia, H., Lynch, R., Wang, Y., & Fox-Rushby, J. (2021). Impact of multi-morbidity on healthcare costs and utilisation: A systematic review of the UK literature. *British Journal of General Practice, 71*(702), e39. https://doi.org/10.3399/bjgp20x713897

Suls, J., Green, P. A., & Boyd, C. M. (2019). Multimorbidity: Implications and directions for health psychology and behavioral medicine. *Health Psychology, 38*(9), 772–782. https://doi.org/10.1037/hea0000762

Switankowsky, I. (2000). Dualism and its Importance for Medicine. *Theoretical Medicine and Bioethics, 21*(6), 567–580. https://doi.org/10.1023/a:1026570907667

Teasdale, T. W., & Engberg, A. W. (2001). Suicide after traumatic brain injury: A population study. *Journal of Neurology, Neurosurgery and Psychiatry, 71*(4), 436. https://doi.org/10.1136/jnnp.71.4.436

du Vaure, C. B., Dechartres, A., Battin, C., Ravaud, P., & Boutron, I. (2016). Exclusion of patients with concomitant chronic conditions in ongoing randomised controlled trials targeting 10 common chronic conditions and registered at ClinicalTrials.gov: A systematic review of registration details. *BMJ Open, 6*(9), e012265. https://doi.org/10.1136/bmjopen-2016-012265

Wang, Y.-P., Nunes, B. P., Coêlho, B. M., Santana, G. L., Nascimento, C. F., & do, Viana, M. C., Benseñor, I. M., Andrade, L. H., and Filho, A. D. P. C. (2019). Multilevel analysis of the patterns of physical-mental multimorbidity in general population of São Paulo metropolitan area, Brazil. *Scientific Reports, 9*(1), 2390. https://doi.org/10.1038/s41598-019-39326-8

Waters, L., Cameron, K., Nelson-Coffey, S. K., Crone, D. L., Lomas, T., Oades, L., Owens, R. L., Pawelski, J. O., Rashid, T., Warren, M. A., White, M. A., & Williams, P. (2021). Collective wellbeing and posttraumatic growth during COVID-19: How positive psychology can help families, schools, workplaces and marginalized communities. *The Journal of Positive Psychology*, Advance online publication. https://doi.org/10.1080/17439760.2021.1940251

Whiteford, H. A., Degenhardt, L., Rehm, J., Baxter, A. J., Ferrari, A. J., Erskine, H. E., Charlson, F. J., Norman, R. E., Flaxman, A. D., Johns, N., Burstein, R., Murray, C. J. L., & Vos, T. (2013). Global burden of disease attributable to mental and substance use disorders: Findings from the global burden of disease study 2010. *Lancet, 382*(9904), 1575–1586. https://doi.org/10.1016/s0140-6736

Whitty, C. J. (2017). Harveian oration 2017: Triumphs and challenges in a world shaped by medicine. *Clinical Medicine, 17*(6), 537–544. https://doi.org/10.7861/clinmedicine.17-6-537

Whitty, C. J. M., MacEwen, C., Goddard, A., Alderson, D., Marshall, M., Calderwood, C., Atherton, F., McBride, M., Atherton, J., Stokes-Lampard, H., Reid, W., Powis, S., & Marx, C. (2020). Rising to the challenge of multimorbidity. *The British Medical Journal, 368*, l6964. https://doi.org/10.1136/bmj.l6964

Wilkie, L., Arroyo, P., Conibeer, H., Kemp, A. H., & Fisher, Z. (2021). The impact of psycho-social interventions on the wellbeing of individuals with acquired brain injury during the COVID-19 pandemic. *Frontiers in Psychology, 12*, 648286. https://doi.org/10.3389/fpsyg.2021.648286

Wilkie, L., Fisher, Z., & Kemp, A. H. (2022). The complex construct of wellbeing and the role of vagal function. *Frontiers in Integrative Neuroscience, 16*, 925664. https://doi.org/10.3389/fnint.2022.925664/full

Willis, C., Riley, B., Herbert, C., Millar, J., & Howland, D. (2014). Systems thinking for transformational change in health. *Evidence and Policy, 10*(1), 113–126. https://doi.org/10.1332/174426413x662815

Wong, P. T. P. (2019). Second wave positive psychology's (PP 2.0) contribution to counselling psychology. *Counselling Psychology Quarterly, 32*, 1–10. https://doi.org/10.1080/09515070.2019.1671320

CHAPTER 5

ACTing for Society: The Promotion and Nurturance of Prosocial Behavior at Scale

Darren J. Edwards

Abstract Acceptance and commitment therapy (ACT) has traditionally been employed at the individual level as a therapeutic model to change the way the individual relates to their thoughts and feelings, helping to connect to meaningful values, and promoting psychological flexibility. More recently, it also has been applied at the societal level to promote prosocial behavior, facilitating shared values within organizations, and promoting a less coercive and more prosocial and equitable society. This has been done by applying evolutionary principles which can guide the selection of individual or group values that are adaptive and promote some advantage to the individual or group (e.g., promoting prosocial connection, compassion, and collective values), providing opportunities for multilevel impacts At an operational level, this work involves applying Nobel prize winner's

D. J. Edwards (✉)
Department of Public Health, School of Health and Social Care, Faculty of Medicine, Health and Life Science, Swansea University, Swansea, Wales, UK
e-mail: D.J.Edwards@Swansea.ac.uk

A. H. Kemp, D. J. Edwards (eds.), *Broadening the Scope of Wellbeing Science*, https://doi.org/10.1007/978-3-031-18329-4_5

Elinor Ostrom's core design principles (CDPs) of organizational psychology, which involves operationalizing at a societal level: (1) group values; (2) equitable solutions; (3) fairness; (4) agreed-upon behaviors; (5) reward; (6) conflict resolution; (7) decentralized; and (8) polycentric governance. This chapter will explain how ACT and Ostrom's CDPs can be applied in practice, whereby they encourage the enabling and the promotion of a more nurturing and prosocial society world which can develop beyond overly individualistic and competitive extrinsic values brought about by neoliberalism, expected utility within a competitive environment, and current Western orientated happiness economics.

Keywords Acceptance and commitment therapy • Multilevel selection • Polyvagal theory • Evolution theory • Prosocial

A major barrier to promoting a nurturing, socially cohesive, and prosocial society is the dominant, hegemonic, economic ideology that 'neoliberalism' places on the development of an individual's values in Western society (i.e., we are more likely to learn neoliberal values from our culture) (Cowden & Singh, 2017; Jaffe & Quark, 2006). Neoliberal policies and values refer to the promotion of individualistic, market-orientated polices such as the deregulation of capital markets, removal of trade barriers, privatization, and the reduction of government regulation over economic activity (Larner, 2003; Thorsen & Lie, 2006; Venugopal, 2015).

Neoliberalism has become embedded in Western culture and is underpinned by a set of social philosophies and worldviews that promote individual self-interest (Thorsen & Lie, 2006). At the same time, neoliberal values deviate away from classical liberalism that emphasizes individual rights (Deleuze & Guattari, 1988), equality and civic obligation (Adams et al., 2019). By contrast, neoliberalism redefines the citizen as a consumer (Payne, 2012), whose choices are exercised by buying and selling through a process that rewards selfish immediate gain, promoting competition as the defining characteristic of human relations in a zero-sum game (i.e., the gain for one individual, can mean loss for another) (Clarke, 2005).

In the documentary *HyperNormalization*, Adam Curtis (2016) refers to a society where the population are aware that governments, technological utopians, and philanthropic financiers have given up on the idea of a complex 'real world', which requires complex solutions to deal with complex problems, and instead have accepted a simpler 'fake world' run by neoliberal ideology in the form of corporations and kept stable by politicians. He suggests that the general population know this, yet apathy leads to inaction. He explains that people feel trapped in this world, which lacks any real societal alternative. Such conclusions are indeed supported by the evidence, which shows that our hypercapitalist, neoliberal consumerist society reinforces a model of material influence as the only path to happiness and a meaningful life (Binkley, 2014; Kasser & Sheldon, 2009; Montes, 2020). The result of such hypercapitalist extrinsic values leads to 'time poverty' (not having enough time due to working long hours) (De Graaf, 2003). A consequence of this materialist and consumer-driven society is that people are working long hours, leading to a reduction in positive affect, more negative affect, and lower life satisfaction (Kasser & Brown, 2003; Kasser & Sheldon, 2009).

A neoliberal hegemonic culture promotes a perception of various social issues and phenomena which are predominately conceptualized through an individualistic lens (Dickie et al., 2006; Madsen et al., 2016). This ideology may be the root of many of our problems including commodification of mental health problems (Esposito & Perez, 2014), income inequality, disempowerment of workers, inadequate social services, ineffective healthcare systems (Zeira, 2022) and even the climate catastrophe (Weintrobe, 2021).

Linked to neoliberalism is the obsession within Western society to chase and capture individual happiness (Burnett, 2011; Gruber et al., 2011; Joshanloo, 2014; McMahon, 2006). Happiness has become the new moral compass in neoliberal societies which defines what is right and wrong and is framed as a model of selfhood which is aligned to individualism and consumerism (Cabanas, 2016). Neoliberal exploitation of the individual whose value within society is only that of a consumer, has led to excessive consumer behavior which only leads to immediate short term fleeting forms of happiness (Giroux, 2011; Lyons, 2015; Tremblay, 2015). Buying a new iPhone, a new car, keeping up with the Jones', are examples of modern obsessions, despite only leading to fleeting experience of happiness. Studies of happiness economics which employ temporal discounting tasks have shown that people tend to prefer immediate and smaller gains than larger ones over the longer term (Green et al., 2005; Van den Bos & McClure, 2013).

Yet, chasing fleeting forms of happiness has been shown to have a paradoxical effect such as increased risk of depression (Ford & Mauss, 2014;

Ford et al., 2014), lower wellbeing, decreased happiness, and compromised social outcomes (Gruber et al., 2011; Mauss et al., 2011; Schooler et al., 2003). The problem with centering one's life and goals around something as elusive as happiness is that this may distract the individual away from meaningful intrinsically value-oriented behaviors.

Further consolidating the problem, and causing confusion, happiness economics attempts to broadly measure wellbeing from both objective and subjective measures, for example, life expectancy, age, education, income, marital status, institutional quality, GDP statistics, and various other arbitrary life satisfaction indices, which can be placed into dimensions of affective, evaluative, and eudaimonic scales (Gielen & Van Ours, 2014; Nikolova & Graham, 2021; O'Connor, 2020). These measures are based on assumptions that do not always align with outcomes of wellbeing nor are a direct measure of it. For example, it assumes that neoliberal growth, e.g., GDP of a nation is equivalent to growth in a nations wellbeing. However, it has long been established that increases in gross domestic product (GDP) beyond a threshold of basic needs do not lead to further increases in wellbeing (Pretty et al., 2016).

The selfish pursuit of individual happiness, narrowly defined may lead to problems at a societal level, represented by neoliberal societies, hypercompetitiveness (e.g., GDP growth, income), and a disregard for connecting to the natural environment, personal connection to others and greater prosocial behaviors. There is a significant difference between internal intrinsic values (values which are deeply personal such as connection, friendship, etc.) and external values driven by societally accepted norms shaped by neoliberalism (such as material gain, possessions, status etc.). Hari (2020) in his book *Lost Connections* makes this point clear, and suggests it is intrinsic values which are vital for achieving wellbeing. Evidence from dozens of studies has converged on the conclusion that people who prioritize extrinsic values of money, image, and status have lower personal wellbeing and higher distress (Jackson, 2005; Kasser, 2003). These extrinsic values, according to Hari (2020) are those which are imposed on us, and are influenced by externally accepted materialistic and neoliberal values. He also suggests that internal intrinsic values are those which really matter to us irrespective of external pressures. This could be something as simple as learning to play the piano, painting some art, connecting with nature, etc. Hari suggests that it is these intrinsic values which are likely to make the individual more fulfilled, increasing their wellbeing.

A Solution Based on Intrinsic Values Identification and Prosocial Behavior

There is hope that people can rise beyond Curtis' (2016) HyperNormalization to build prosocial and cooperative values, connect to other people and foster a deeper concern for the environment, and the Earth we live on. Yuval Noah Harari (2014) suggests in his book *Sapiens: A Brief History of Humankind*, that humans can imagine through language, a new world, and new alternatives. Humans have a remarkable capacity to think of new ideas and this can include a new society, and a new way of living. This can be done in two ways, either through a radical shift in society, or from the ground up, engaging with individual's personal values, promoting prosocial behavior, and fostering growth towards more cooperative behavior and less competitive, coercive, and divisive behavior.

One psychological therapeutic approach which may help the individual identify and engage with such intrinsic values, which Hari (2020) discusses, is called acceptance and commitment therapy (ACT) (Hayes et al., 2009, 2011). This model promotes psychological flexibility through the engagement of personally meaningful intrinsic values. Psychological flexibility can be defined as a process of engaging with openness to pain, acceptance, awareness of thoughts and feelings which unfold in the present moment, as well as changing or persisting in behavior which is aligned to the individuals' central goals and values (Bond et al., 2006; Kashdan & Rottenberg, 2010). ACT has been found to promote wellbeing (Fox, 2021; Gloster et al., 2020; Katajavuori et al., 2021), and has recently been applied to evolutionary principles in order to realized collective wellbeing. This distinguishes the individual-level values, and behavioral selection to that of the group level. As such, it has been recognized that at the group level; we, as individuals, also need to fulfill complex and prosocial needs and not just individual ones, by selecting values and behaviors at this group level which promote co-operation within groups and communities, such as basic human yearnings and need for connection, love, and friendship (Atkins et al., 2019; Wilson & Sober, 1994). This is distinct from the neo-liberal worldview of selfish individual gain and competitiveness. There should be greater variability in choices for those who are psychologically flexible, whilst those who are less psychologically flexible display a more rigid set of behaviors.

Values, goals, emotions, and cognition all have a dimension of variability and have been accounted for within the extended evolutionary meta model (EEMM) (Hayes, Hofmann, & Ciarrochi, 2020). In this model, a process of change is a contextually situated, modifiable behavior or sequence of behaviors that orients the individual towards an adaptive outcome. Evolutionary principles in this regard revolve around selection, i.e., identifying the success of those values in maintaining the values of the individual or the group (e.g., compassion, prosocial efforts over profit, or community connection over individual selfish pursuit—at the sociocultural level), much in the same way as biological evolution selects biological features that lead to a survival advantage of the organism. This, therefore, frames selection of values whether individual or group as a fitness function for the promotion of wellbeing, such as the section of prosociality, social cohesion, compassion, and nature connectedness, all of which are intrinsic, non-coercive, and extend to the group and community level.

Biologically, evolution has facilitated the separation of humans from other mammals, enabling key aspects relating to socialization such as complex language to be developed. Socialization is also supported by the autonomic nervous system function functioning, as described in polyvagal theory (PVT) (Porges, 1995, 2018), such that the most evolutionary recent ventral myelinated aspect of the vagus nerve supports affiliative behaviour (Porges, 2007). The vagus nerve innervates a number of organs which have a role to play in communication and emotion (Porges, 1995, 2022).

When an individual feels unsafe and threatened by their environment, such as when experiencing hyper-competitiveness, lack of compassion and connection, associated with, for example, the extrinsic values a neoliberal culture brings about, then the vagal break is withdrawn, providing permission for the sympathetic nervous system to dominate a defensive reaction (Porges, 2009, 2022), giving rise to the fight-or-flight system and increased activity of the hypothalamic pituitary axis (HPA) (allowing increased cortisol and adrenaline). This leads to more rigid and less adaptive behavior, which is protectionist, and discourages social connection which then inhibits group-level prosocial values. However, when engaging in psychological flexibility, behavior is more adaptive (Gloster et al., 2017), promotes social connection and prosocial behavior (Gloster et al., 2020), inhibits the HPA-related stress and trauma (Richardson & Jost, 2019), thus highlighting the benefits of promoting adaptive vagal functioning (i.e., by activating the calming function of the vagal break), and improving positive health and wellbeing (Kashdan & Rottenberg, 2010) supported by prosociality. This psychological flexibility, and supported by adaptive vagal functioning, may facilitate greater selection of prosocial values at a group level.

More broadly, ACT combined with evolutionary adaptation theory has recently developed within a framework called prosocial, which utilizes the ACT matrix within an evolutionary approach (Atkins et al., 2019; Wilson & Sober, 1994), which is readily applicable to theories of vagal functioning. Evolutionary principles of selection are traditionally focused on the individual (Cosmides & Tooby, 2000), but in the case of prosocial, ACT extends with evolutionary principles to allow it to practically work within a societal context, as it can focus on both individual- and group-level selection, thus promoting multilevel selection (Wilson & Sober, 1994). In this framework, selection is made on behavioral variation, including individual and group-based values, supporting prosocial behavior i.e., through concretely selecting values at the level needed to achieve greater prosociality.

Within this 'prosocial' framework, multilevel selection is guided by the application of Elinor Ostrom's core design principles (CPDs), for which she won a Nobel prize in economics. Ostrom and colleagues work (Ostrom, 1990, 2010, 2019; Ostrom & Cox, 2010) focused on how humans interact with ecosystems in order to maintain long-term sustainability. She explored how societies have developed diverse institutional arrangements for managing natural resources and avoiding economic collapse. As part of this work, she developed an evidence base for a more general theory of individual choice that recognized a central role of trust and cooperation which she described as common-pool resources (called the CDPs) for optimal self-organization.

Specifically, the CPDs applied to prosocial (Atkins et al., 2019; Wilson et al., 2020), relate to the selection of (1) group values, to help the group to identify their purpose; (2) equitable solutions within the group, to ensure equitable contribution and benefits between members, and ensuring there are variable and multiple perspectives from the group; (3) fairness, in ensuring decision-making is fair and inclusive; (4) agreed-upon behaviors which are monitored by group members; (5) reward, by ensuring helpful behaviors are rewarded, and unhelpful behaviors are put into extinction; (6) conflict resolution, where conflicts are resolved quickly and fairly; (7) decentralized governance, by allowing the authority of members and sub-groups to self-govern; and (8) polycentric governance, whereby the group utilize principles 1–7 to relate and collaborate with other groups. Application of the 'prosocial' framework to the CPDs have shown they can improve further cooperation, minimize coercion, and increase equity across members of different groups (Atkins et al., 2019; Wilson et al., 2020). Cooperation in small groups has been improved due to engaging with the 'prosocial' framework (Hayes et al., 2021), as well as accelerating leadership skills whilst reducing burnout, stress, and injuries (Moran et al., 2021). Though these developments are quite recent, they clearly have promising applications for promoting prosocial behavior at both the individual and societal level.

Conclusion

Within popular culture, an apparent desire to desperately escape the confinements of neoliberalism can easily be found. One example is through the blockbuster movie *The Matrix* (Wachowskis, 1999), which has been cited as a postmodern allegory for the contemporary commercialized, media-driven, neoliberal, and postmodern society (Alen, 2012), which adapts the philosophical treatise *Simulacra and Simulation* by sociologist Jean Baudrillard (1994) about the relation between reality, symbols, and society, and an almost unconscious motivation to escape the neoliberal confinements of society. However, popular culture does not provide realistic solutions to reform society; instead, it offers a dream. We cannot consume a pill to escape the confinements of big tech and mass industrialization brought about by neoliberalism like in the movie *The Matrix*.

However, ACT applied through evolutionary principles and the prosocial framework of Ostrom's principles may have benefits for helping the individual connect to intrinsic values and, at the group level, supporting social cohesion and connectedness, which may help overcome entrenched external neoliberal societal values relating to immediate self-interest. Prosocial (Atkins et al., 2019) can encourage compassion for each other and social connection, supporting the identification of shared goals, values and community. This framework challenges the societal disconnection associated with individualism at an individual level and neoliberalism at the societal level. The focus for prosocial is on promoting more productive and cooperative behaviour, and less coercive manipulation, facilitated through the application of Elinor Ostrom's core design principles to the organization of groups (Atkins et al., 2019; Wilson et al., 2020). The benefits may even extend to greater ecological sustainability such as environmental care and action to avoid climate change.

References

Adams, G., Estrada-Villalta, S., Sullivan, D., & Markus, H. R. (2019). The psychology of neoliberalism and the neoliberalism of psychology. *Journal of Social Issues, 75*(1), 189–216.

Alen, J. (2012). The matrix and postmodernism. Avialiable at: https://prezi.com/ybxwvr21r9lz/the-matrix-andpostmodernism/

Atkins, P. W., Wilson, D. S., & Hayes, S. C. (2019). *Prosocial: Using evolutionary science to build productive, equitable, and collaborative groups.* New Harbinger Publications.

Baudrillard, J. (1994). *Simulacra and simulation.* University of Michigan Press.

Binkley, S. (2014). *Happiness as enterprise: An essay on neoliberal life.* Suny Press.

Bond, F. W., Hayes, S. C., & Barnes-Holmes, D. (2006). Psychological flexibility, ACT, and organizational behavior. *Journal of Organizational Behavior Management, 26*(1–2), 25–54.

Burnett, S. (2011). *The happiness agenda: A modern obsession.* Springer.

Cabanas, E. (2016). Rekindling individualism, consuming emotions: Constructing "psytizens" in the age of happiness. *Culture & Psychology, 22*(3), 467–480.

Clarke, S. (2005). The neoliberal theory of society. *Neoliberalism: A critical reader, 50*, 59.

Cosmides, L., & Tooby, J. (2000). Evolutionary psychology and the emotions. In M. Lewis, & J. M. Haviland-Jones (Eds.), *Handbook of Emotions* (Vol. 2, Iss. 2, pp. 91–115), Guilford Press.

Cowden, S., & Singh, G. (2017). Community cohesion, communitarianism and neoliberalism. *Critical Social Policy, 37*(2), 268–286. https://doi.org/10.1177/0261018316670252

Curtis, A. (2016). HyperNormalisation. *BBC Documentary.* Available at: https://www.youtube.com/watch?v=-fny99f8amM

De Graaf, J. (2003). *Take back your time: Fighting overwork and time poverty in America.* Berrett-Koehler Publishers.

Deleuze, G., & Guattari, F. (1988). *A thousand plateaus: Capitalism and schizophrenia.* Bloomsbury Publishing.

Dickie, V., Cutchin, M. P., & Humphry, R. (2006). Occupation as transactional experience: A critique of individualism in occupational science. *Journal of Occupational Science, 13*(1), 83–93.

Esposito, L., & Perez, F. M. (2014). Neoliberalism and the commodification of mental health. *Humanity and Society, 38*(4), 414–442.

Ford, B., & Mauss, I. (2014). The paradoxical effects of pursuing positive emotion: When and why wanting to feel happy backfires. In J. Gruber & J. T. Moskowitz (Eds.), *Positive Emotion: Integrating the Light Sides and Dark Sides* (pp. 363–382). Oxford Academic.

Ford, B., Shallcross, A. J., Mauss, I. B., Floerke, V. A., & Gruber, J. (2014). Desperately seeking happiness: Valuing happiness is associated with symptoms and diagnosis of depression. *Journal of Social and Clinical Psychology, 33*(10), 890.
Fox, J. (2021). A reflection on the value of acceptance and commitment therapy for promoting good mental health in the workplace: A service user perspective. *British Journal of Guidance & Counselling, 49*(3), 362–370.
Gielen, A. C., & Van Ours, J. C. (2014). Unhappiness and job finding. *Economica, 81*(323), 544–565.
Giroux, H. A. (2011). Neoliberalism and the death of the social state: Remembering Walter Benjamin's angel of history. *Social Identities, 17*(4), 587–601.
Gloster, A. T., Meyer, A. H., & Lieb, R. (2017). Psychological flexibility as a malleable public health target: Evidence from a representative sample. *Journal of Contextual Behavioral Science, 6*(2), 166–171.
Gloster, A. T., Rinner, M. T., & Meyer, A. H. (2020). Increasing prosocial behavior and decreasing selfishness in the lab and everyday life. *Scientific Reports, 10*(1), 1–9.
Green, L., Myerson, J., & Macaux, E. W. (2005). Temporal discounting when the choice is between two delayed rewards. *Journal of Experimental Psychology: Learning, Memory, and Cognition, 31*(5), 1121.
Gruber, J., Mauss, I. B., & Tamir, M. (2011). A dark side of happiness? How, when, and why happiness is not always good. *Perspectives on Psychological Science, 6*(3), 222–233.
Harari, Y. N. (2014). *Sapiens: A brief history of humankind.* Random House.
Hari, J. (2020). *Lost connections.* Bloomsbury Publishing.
Hayes, S. C., Strosahl, K. D., & Wilson, K. G. (2009). *Acceptance and commitment therapy.* American Psychological Association.
Hayes, S. C., Strosahl, K. D., & Wilson, K. G. (2011). *Acceptance and commitment therapy: The process and practice of mindful change.* Guilford Press.
Hayes, S. C., Hofmann, S. G., & Ciarrochi, J. (2020). A process-based approach to psychological diagnosis and treatment: The conceptual and treatment utility of an extended evolutionary meta model. *Clinical Psychology Review, 82*, 101908.
Hayes, S. C., Atkins, P., & Wilson, D. S. (2021). Prosocial: Using an evolutionary approach to modify cooperation in small groups. In R. A. Houmanfar, M. Fryling, & M. P. Alavosius (Eds.), *Applied Behavior Science in Organizations* (pp. 197–223). Routledge.
Jaffe, J., & Quark, A. A. (2006). Social cohesion, neoliberalism, and the entrepreneurial community in rural Saskatchewan. *American Behavioral Scientist, 50*(2), 206–225. https://doi.org/10.1177/0002764206290634
Jackson, T. (2005). Live better by consuming less?: is there a "double dividend" in sustainable consumption?. *Journal of Industrial Ecology, 9*(1–2), 19–36.
Joshanloo, M. (2014). Eastern conceptualizations of happiness: Fundamental differences with western views. *Journal of Happiness Studies, 15*(2), 475–493.

Kashdan, T. B., & Rottenberg, J. (2010). Psychological flexibility as a fundamental aspect of health. *Clinical Psychology Review, 30*(7), 865–878.

Kasser, T. (2003). *The high price of materialism*. MIT press.

Kasser, T., & Brown, K. W. (2003). On time, happiness, and ecological footprints. In J. de Graaf (Ed.), *Take Back Your Time: Fighting Overwork and Time Poverty in America* (pp. 107–112). Berrett-Koehler Publishers.

Kasser, T., & Sheldon, K. M. (2009). Time affluence as a path toward personal happiness and ethical business practice: Empirical evidence from four studies. *Journal of Business Ethics, 84*(2), 243–255.

Katajavuori, N., Vehkalahti, K., & Asikainen, H. (2021). Promoting university students' well-being and studying with an acceptance and commitment therapy (ACT)-based intervention. *Current Psychology*, 1–13.

Larner, W. (2003). Neoliberalism? *Environment and Planning D: Society and Space, 21*(5), 509–512. https://doi.org/10.1068/d2105ed

Lyons, P. (2015). *The re-enchantment of the world: The value of spirit against industrial populism*. Bloomsbury.

Madsen, J., Marie Kanstrup, A., & Josephsson, S. (2016). The assumed relation between occupation and inequality in health. *Scandinavian Journal of Occupational Therapy, 23*(1), 1–12.

Mauss, I. B., Tamir, M., Anderson, C. L., & Savino, N. S. (2011). Can seeking happiness make people unhappy? Paradoxical effects of valuing happiness. *Emotion, 11*(4), 807.

McMahon, D. M. (2006). *Happiness: A history*. Grove Press.

Montes, J. (2020). Neoliberal environmentality in the land of Gross National Happiness. *Environment and Planning E: Nature and Space, 3*(2), 300–322.

Moran, D. J., Batten, S., Gamble, M. A., & Atkins, P. (2021). Acceptance and commitment training: Improving performance in organizations with applied contextual behavioral science. Applied behavior science in organizations (pp. 177–196). Routledge.

Nikolova, M., & Graham, C. (2021). The economics of happiness. In K. F. Zimmermann (Ed.), *Handbook of Labor, Human Resources and Population Economics* (pp. 1–33). Springer.

O'Connor, K. J. (2020). Life satisfaction and noncognitive skills: Effects on the likelihood of unemployment. *Kyklos, 73*(4), 568–604.

Ostrom, E. (1990). *Governing the commons: The evolution of institutions for collective action*. Cambridge University Press.

Ostrom, E. (2010). Beyond markets and states: Polycentric governance of complex economic systems. *American Economic Review, 100*(3), 641–672.

Ostrom, E. (2019). Institutional rational choice: An assessment of the institutional analysis and development framework. In P. A. Sabatier (Ed.), *Theories of the Policy Process* (pp. 21–64). Routledge.

Ostrom, E., & Cox, M. (2010). Moving beyond panaceas: A multi-tiered diagnostic approach for social-ecological analysis. *Environmental Conservation, 37*(4), 451–463.

Payne, C. (2012). *The consumer, credit and neoliberalism: Governing the modern economy*. Routledge.
Porges, S. W. (1995). Orienting in a defensive world: Mammalian modifications of our evolutionary heritage. A polyvagal theory. *Psychophysiology, 32*(4), 301–318.
Porges, S. W. (2007). The polyvagal perspective. *Biological Psychology, 74*(2), 116–143. https://doi.org/10.1016/j.biopsycho.2006.06.009
Porges, S. W. (2009). The polyvagal theory: New insights into adaptive reactions of the autonomic nervous system. *Cleveland Clinic Journal of Medicine, 76*(Suppl 2), S86.
Porges, S. W. (2018). Polyvagal theory: A primer. In S. W. Porges (Ed.), *Clinical Applications of the Polyvagal Theory: The Emergence of Polyvagal-Informed Therapies* (Vol. 50, p. 69). W. W. Norton & Company.
Porges, S. W. (2022). Polyvagal theory: A science of safety. *Frontiers in Integrative Neuroscience, 16*, 871227.
Pretty, J., Barton, J., Pervez Bharucha, Z., Bragg, R., Pencheon, D., Wood, C., & Depledge, M. H. (2016). Improving health and well-being independently of GDP: Dividends of greener and prosocial economies. *International Journal of Environmental Health Research, 26*(1), 11–36.
Richardson, C. M., & Jost, S. A. (2019). Psychological flexibility as a mediator of the association between early life trauma and psychological symptoms. *Personality and Individual Differences, 141*, 101–106.
Schooler, J., Ariely, D., & Loewenstein, G. (2003). The pursuit and assessment of happiness can be self-defeating. In I. Brochas & J. Carrillo (Eds.), *The Psychology of Economic Decisions* (p. 1). Oxford University Press.
Thorsen, D. E., & Lie, A. (2006). *What is neoliberalism* (pp. 1–21). University of Oslo, Department of Political Science. Manuscript.
Tremblay, T. (2015). *Reading in an age of neoliberalism* (Vol. 32, p. 312). University of Toronto Press.
Van den Bos, W., & McClure, S. M. (2013). Towards a general model of temporal discounting. *Journal of the Experimental Analysis of Behavior, 99*(1), 58–73.
Venugopal, R. (2015). Neoliberalism as concept. *Economy and Society, 44*(2), 165–187.
Wachowskis, T. (1999). *The Matrix W*. Warner Bros.
Weintrobe, S. (2021). *Psychological roots of the climate crisis: Neoliberal exceptionalism and the culture of uncare*. Bloomsbury Publishing.
Wilson, D. S., & Sober, E. (1994). Reintroducing group selection to the human behavioral sciences. *Behavioral and Brain Sciences, 17*(4), 585–608.
Wilson, D. S., Philip, M. M., MacDonald, I. F., Atkins, P. W., & Kniffin, K. M. (2020). Core design principles for nurturing organization-level selection. *Scientific Reports, 10*(1), 1–6.
Zeira, A. (2022). Mental health challenges related to neoliberal capitalism in the United States. *Community Mental Health Journal, 58*(2), 205–212.

CHAPTER 6

The Relationship Between Nature Connectedness and Human and Planetary Wellbeing: Implications for Promoting Wellbeing, Tackling Anthropogenic Climate Change and Overcoming Biodiversity Loss

Alison Pritchard and Miles Richardson

Abstract Human wellbeing is inextricably linked with nature's wellbeing. If we harm our natural environment then we harm ourselves, and the risks from loss of habitats and a warming planet are incalculable.

The climate crisis and wildlife emergency show that the existing relationship between people and the rest of nature is failing. Too often we see nature as something to use, to control or as a threat to us, rather than as a fundamental part of ourselves. We urgently need a new relationship with nature in order to transform our attitudes and behaviours into positive and

A. Pritchard (✉) • M. Richardson
School of Psychology, University of Derby, Derby, UK
e-mail: A.Pritchard@derby.ac.uk; M.Richardson@derby.ac.uk

A. H. Kemp, D. J. Edwards (eds.), *Broadening the Scope of Wellbeing Science*, https://doi.org/10.1007/978-3-031-18329-4_6

meaningful outcomes for the environment, and help tackle the crisis in our mental health and wellbeing.

Recent research evidence suggests that being connected to nature is not an optional extra for achieving good health and wellbeing, but a basic human need. We must build a new relationship by developing an affinity with nature and by celebrating the role of nature in healthy, sustainable and meaningful lives.

This chapter explores theories of nature connectedness, wellbeing and pro-environmental behaviours, the contexts in which people become related to nature, the role that different types and qualities of nature have in this relationship, and the key societal and individual factors necessary for strengthening this connection. We call for a life-course approach towards developing and nurturing our affinity with nature as a vital means of safeguarding both human and planetary well-being.

Keywords Nature connectedness • Wellbeing • Pro-environmental behaviour • Climate change • Biodiversity

Introduction

> At the outset of the 21st century there is no longer any doubt that a strong individual sense of connection to nature and natural processes is utterly essential to the healthy coexistence of humans with their biological neighbours and physical setting. (Pyle, 2003, p. 206).

Human behaviour (or misbehaviour) has been a fundamental cause of the major environmental crises—climate change, land degradation and loss of biodiversity—that have devastated our natural world and threaten our future survival (Rabinowitz et al., 2018; Ripple et al. 2020; Wiedmann et al., 2020). Such changes are implicated as both cause and consequence of poor human wellbeing, and declining wellbeing and growing rates of mental ill-health are cause for grave concern world-wide (Rabinowitz et al., 2018). The fact that nature connectedness (NC) is increasingly recognised as a crucial aspect of human wellbeing (Lambert et al., 2020; Mead et al., 2021; White et al., 2021) leads us to explore the role that a close connection with nature may have in providing urgently needed solutions.

Nature Connectedness

NC is the psychological construct that describes an individuals' relationship with nature. It incorporates feelings and emotions, perceptions of nature and how it relates to the sense of self, and ways of living and being in nature. At an individual level, there is ample evidence that NC is associated with a broad range of hedonic (feeling) and eudaimonic (functioning) wellbeing benefits such as happiness, meaning and purpose in life, vitality, personal growth and autonomy (Capaldi et al., 2014; Pritchard et al., 2019). NC is a significant and distinct predictor of wellbeing over and above the wellbeing benefits associated with human social connectedness (Lambert et al., 2020), and NC is also a stronger predictor of eudaimonic wellbeing than is socio-economic status (Martin et al., 2020).

Importantly, the benefits of NC do not arise from us simply spending more *time* in nature, but, rather, relate directly to our personal *engagement* with nature, and to the depth and quality of our experiences (Richardson et al., 2021; Zylstra et al., 2014). NC is more than just a transient 'warm glow' feeling after spending time in nature; rather it is 'an enduring appreciation, empathy, and mindfulness of the intrinsic value and shared essence of all life' which includes: 'an appreciative understanding of the interconnectedness of life as well as experiences and behaviours that exhibit action and agency toward the environment' (Zylstra et al., 2014, p. 126).

The concept of NC is linked to the idea of an 'ecological self' which refers to an expansive sense of self which includes all life-forms, ecosystems and the Earth itself (Bragg, 1996). Nature is incorporated into an individuals' self-definition and includes both affective and cognitive aspects; these centre on our identity being a part of nature, our emotional bond with nature and our feelings of love and care for the environment.

Our varied emotional responses to nature are also important factors in the link between NC and wellbeing. We react with awe and wonder, with joy and happiness, with contentedness, with fascination and, sometimes, with fear (Capaldi et al., 2014; Shiota et al., 2007). Individuals with high levels of NC are more likely to report feelings of awe and other self-transcendent experiences (e.g. Davis & Gatersleben, 2013), and nature-induced awe leads to increases in wellbeing (Anderson et al., 2018). NC also has a causal role in mediating the relationship between awe and ecological behaviour (Yang et al., 2018).

And there is growing evidence that the natural world, and our connection with it, profoundly influences how we grow and develop—emotionally, cognitively and spiritually (Barton et al., 2016; Kahn, 1997). The wellbeing importance of NC is closely associated with the presence of meaning (Howell et al., 2013), spirituality (Trigwell, Francis, & Bagot, 2014) and self-transcendence (Davis & Gatersleben, 2013). Such immersive experiences are especially felt in wilderness areas (Barton et al., 2016; Davis & Gatersleben, 2013) and in situations where natural phenomena are vast or unexpected in size or scope (Yang et al., 2018). We can be overwhelmed by a sense of our own smallness, but this reminds us of the interconnectedness of all things and gives the impetus to rethink our values and worldviews. Intense experiences of awe and self-transcendence, felt through being deeply immersed in nature, are associated with altruism and pro-social behaviour (Castelo et al., 2021), and may be important in the development of wisdom (Keltner & Piff, 2020).

NC is increasingly viewed as a basic psychological need (Baxter & Pelletier, 2018), but we also need to recognise that there is an associated cost: as with any relationship there is an inevitable sense of grief and loss that accompanies experiences of current ecological degradation and biodiversity loss (Chawla, 2020). The existence of emotional and cultural attachment to a specifical geographical location (topophilia) has close links with nature affiliation (Beery et al., 2015), and Riechers et al. (2021) describe how in some rural communities, local inhabitants' awareness of environmental degradation has damaged their emotional connection to the land. Place attachment is strongly associated with NC, and mediates the relationship between NC and wellbeing (Basu et al., 2021). To counter the sense of dissociation and loss, Riechers et al. (2021) highlight the importance of NC as a means of fostering feelings of individual agency—inspiring people to take action to protect the environment. In this way NC helps build 'constructive hope'—the capacity to face environmental threats and uncertainties, while finding positive meaning in taking action (Chawla, 2020).

Our emotional attachment to nature leads to an expanded sense of self and increased empathy with, and concern for, non-human nature; this in turn creates a greater desire to protect it (Mackay & Schmitt, 2019). NC is strongly associated with a range of pro-environmental behaviours (including pro-biodiversity conservation behaviours); this includes evidence for a causative link between NC and PEB (Mackay & Schmitt, 2019). This relationship seems to be mutually reinforcing—there is growing evidence that biological and habitat diversity are, in turn, linked to human psychological wellbeing (e.g. Aerts et al., 2018; WHO, 2015), with NC being an important factor in this association (Wyles et al., 2019).

Building an individual affinity with nature seems to be especially important for the harder-to-engage-in types of environmental behaviours such as pro-biodiversity conservation behaviours (Mackay & Schmitt, 2019; Martin et al., 2020), and there are increased calls to recognise the importance of relational values in guiding the design and implementation of pro-biodiversity conservation measures (Chan et al., 2016; Richardson, Passmore, et al., 2020).

Extinction of Experience

'Extinction of experience' describes how humans over time have become distanced—even estranged—from nature, and the consequent diminution of human health and wellbeing benefits associated with the natural world (Miller, 2005). Orr (1993, p. 417) described this process as a 'slow tectonic shift in perception and attitudes that widened throughout the Middle Ages to the present'. The reasons for this alienation include societal factors such as increased industrialisation and urbanisation, technological advancements, and the dominance of individualistic and anthropocentric values (Miller, 2005); as well as individual behavioural factors such as the growth of sedentary lifestyles and pastimes; increased use of social media; more time spent indoors; and the loss of free play opportunities in childhood (Chawla, 2020; Soga & Gaston, 2006). And there is evidence that children from minority ethnic backgrounds or from poorer households are least likely to access nature, further exacerbating health and wellbeing inequalities (Mental Health Foundation, 2021).

Allied to the extinction of experience is the problem of 'shifting-baseline syndrome' whereby people judge the health of an environment against the 'best' state of nature that they remember from childhood (Miller, 2005; Pyle, 2003), regardless of whether their childhood memories were of an already impoverished landscape. Extinction of experience results in a negative feedback loop of accelerating disconnection from nature (Soga & Gaston, 2006); this in turn has resulted in a wholesale loss of common knowledge (Pyle, 2003).

As our daily lives have become almost entirely divorced from the need to be in direct contact with nature, so the human mindset has increasingly solidified into a narrow-minded and convenient placidity. We are intolerant of nature and ignorant of natural processes because it suits us to be so. But at the same time we have not lost our sense of superiority, or our habits of dominance and control.

Nature Re-connection

We need concerted action to heal and re-build our relationship with nature for both human and environmental wellbeing. Noticing nature and engaging more deeply are known to strengthen our individual relationship with nature (Passmore & Holder, 2017; Richardson & Sheffield, 2017) and improve mental wellbeing (McEwan et al., 2019). Richardson, Dobson, et al. (2020) have highlighted a number of factors that help build this connection: being aware of nature's beauty, making use of all our senses to notice nature, caring and taking action, finding meaning in nature, and our emotions. By utilising these pathways for nature connection, Richardson, Dobson, et al. (2020) highlight real possibilities for individual change in NC and for disseminating those changes more widely via various agencies and organisations into broader society. For example, the UK National Trust have incorporated these factors into the design of their visitor experiences to make them more immersive and engaging, and this has led to measurable growth in the NC levels of National Trust visitors (Richardson, Dobson, et al., 2020).

Building a connection with nature needs to start in childhood if it is to bear most fruit across the lifespan of that individual. Positive childhood experiences in the natural environment are an important factor in adult NC (Windhorst & Williams, 2015), and are known to motivate conservation activities in adulthood (Chawla, 2020). Parental beliefs and attitudes such as talking positively about nature, encouraging children to experience nature outdoors, and being nature connected and environmentally responsible themselves are all important for promoting children's NC (Chawla, 2020; Passmore et al., 2021). Children require safe and accessible places to experience and learn about the natural world, including wilder areas where they have freedom to explore, encounter mystery and stretch their imagination (Chawla, 2020).

Orr (1993, p. 427, 428) has suggested that there may be a critical phase in childhood during which a healthy relationship with nature is forged: 'the capacity for love of any kind begins early in the life and imagination of the child'. If this phase is missed then 'an opportunity will have passed and thereafter the mind will lack some critical dimension of perception and imagination'. Other authors have likewise suggested that childhood NC is a crucial part of healthy child development (Chawla, 2020; Kahn, 1997).

We also need to more fully recognise and value immersive experiences in nature as sources of informal learning (as opposed to formal education) (Barton et al., 2016; Carson, 1956; Soga & Gaston, 2006). It is in our childhood, with the child's intuitive sense of wonder, that we begin our exploration of the world around us; we acquire our reverence for nature, not through being taught, but by freely engaging our senses. As Carson (1956, p. 49, 50) wrote 'I sincerely believe that for the child....it is not half so important to *know* as to *feel*. Once the emotions have been aroused....then we wish for knowledge about the object of our emotional response. Once found it has lasting meaning.'

A Nature Connected Society

Our interest in, and understanding of, NC has grown considerably over the past few years, but it is notable that whilst many studies have focussed on NC at an individual level, there are far fewer that have adopted a collective (societal) perspective (Ives et al., 2017). Whilst is it maybe the case that strengthening *individual* nature connection increases wellbeing and pro-environmental behaviours, how are we to judge the worth of individual change against the inflexible backdrop of overwhelmingly disconnected societies? Real depth of contact—at a societal level—seeks to understand and alter the very value systems and worldviews which are the underlying causes of the ecological crisis (Bragg, 1996; Pyle, 2003). The need to change both our individual and collective relationship with the natural world is paramount (Bragg, 1996; Ives et al., 2017).

The fact that disconnection with nature is perhaps most entrenched in urbanised Western societies has led to calls to look to more traditional ways of life for clues on how to heal this divide, and to recognise the inherent wisdom of such cultures (Nelson, 1993). For inspiration, Nelson (1993) drew on his experiences with the Koyukon people who live as hunter-gatherers in the wildlands of Alaska. For the Koyukons, the material and spiritual aspects of nature are of equal importance in their daily lives. Their 'awareness of the interweaving of human and nonhuman life runs so deep that distinctions between the two seem to fade. There is but one living community in which all organisms—including humankind—hold full and functional membership' (Nelson, 1993, p. 215). The

Koyukons have a profound respect and reverence for nature which manifests itself as a strong community ethic for living within their means and minimising waste.

And change is possible even in countries which are most heavily steeped in agricultural intensification. In her ground-breaking book *Wilding*, Isabella Tree describes the transformation of her 3500-acre Sussex estate in the UK (Knepp Castle) from an intensively managed, species-depleted farm to a thriving patch of land literally blooming with wildlife (Tree, 2018). This was achieved by adopting old-fashioned grazing practices with traditionally bred livestock, and by stopping the use of intensive farming methods such as artificial pesticides and herbicides. But as well as changing their relationship with the land, it was also necessary to build a new relationship with their local community. Although there was initial concern on the part of local residents, and lack of support from established conservation agencies, over time, and with evidence of dramatic increases in biodiversity, improved soil health and thriving landscapes, more widespread acceptance and understanding grew (Tree, 2018).

So What?

Three things seem apparent here. First, we need to relinquish the human tendency, built on domination and utilitarian relationships, to over-control our environment. Letting go of this control is not abdicating responsibility—it is, rather, a recognition of our own limits and of our failures of imagination. And it is an acknowledgement that we don't always have the answers ourselves and that we can learn from what nature teaches us. Second, we need to acknowledge the deep fault-lines in our understanding of what we actually mean by a normal, healthy natural environment. The UK, as well as having among much lower rates of wellbeing and NC than many other Westernised countries (White et al., 2021), is also one of the most bio-depleted countries in the world—this is in itself a tragedy. But alongside this is the equally painful tragedy that we have no memory of the riches we once had.

And third we need to combine our imaginative hope for nature with a thoroughly grounded pragmatism about what can realistically be achieved. Going back is not possible. As Tree observes: 'Our landscape … is so heavily influenced by human beings, so altered by its history and the prevailing conditions of the present, we could only ever hope to create something for the future out of the ingredients remaining to us' (Tree, 2018, p. 153).

The above example gives us hope because it shows how mindsets can be changed and how a completely different way of life can replace an old one in a community. But it is one small example from one small part of the UK. Changing individual NC leads to improved life-long wellbeing and more sustained environmentally positive behaviours. Changing our collective relationship with the land, and becoming more attuned to natural processes, leads to more sustainable land use and farming practices, and stronger, healthier communities. But much more profound societal changes are needed to truly safeguard human and environmental wellbeing into the future.

The value that we place on growth and consumption and the belief that technological progress enables us to overcome the limitations imposed by nature are seen by many as being incompatible with true sustainability and a cause, as well as a consequence, of human disconnection from nature (Pyle, 2003; Zylstra et al., 2014; Martin et al., 2016). It is our profound ignorance of how ecosystems work—our ecological illiteracy—combined with a lack of an intrinsic love and reverence for nature, that has led us to acquiesce in flawed political and economic systems based on infinite growth (Pyle, 2003).

There are increasingly urgent calls for a world-wide reassessment of the damaging impact of growth-oriented policies and affluent overconsumption on the wellbeing of both nature and humans (e.g. Wiedmann et al., 2020). And Martin et al. (2016, p. 6105) highlight the need for a fundamental change in societal values from 'a growth-centred society to one acknowledging biophysical limits and centred on human wellbeing and biodiversity conservation'.

So it is not enough to only make the required relationship changes within *current* societal structures; this does not achieve the fundamental societal shifts necessary to safeguard human and environmental wellbeing into the future (e.g. Orr, 1993; Pyle, 2003). We require a fundamentally new paradigm for a fundamentally nature connected society—a transformative societal change that 'deepens our loyalty and affections to life and in time alters the character of our entire civilisation' (Orr, 1993, p. 430).

Next Steps

The words at the beginning of this chapter were written nearly 20 years ago, in an article in which Pyle set out his ideas for re-connecting humans and nature, but in which he also called his own vision naïve and a 'fairy tale' because it lacked 'any corrective to greed and power' (Pyle, 2003, p. 213). He concluded that 'I have no idea what sort of authority could possibly bring it about.... However, if such a vision were to be....adopted.... as a blueprint by organisations and various levels of authority, I can imagine that slow movement in this direction may be possible' (Pyle, 2003, p. 213).

Since that time, our knowledge of NC has grown considerably. Human values and relatedness have become much more accepted aspects of sustainability science and conservation practice (Chan et al., 2016; Ives et al., 2017; Richardson, Passmore, et al., 2020). And the importance of NC for wellbeing is beginning to be more widely recognised in national and international policy documents (Dasgupta, 2021; WHO, 2015) and wellbeing measures (Lambert et al., 2020).

We can start to re-connect by building strategic NC goals into governmental and organisational policies, increasing opportunities for meaningful interaction with nature close to the places where people live and work, enabling children to create life-long relationships with nature, working to reduce inequalities in access to nature, changing our relationship with the land and focussing on improving biodiversity as well as tackling climate change (Chawla, 2020; Mental Health Foundation, 2021; Richardson, Dobson, et al., 2020; Richardson, Passmore, et al., 2020; Riechers et al., 2021).

How much further, though, we need to go. Although calls for more transformative societal change are growing louder (e.g. Wiedmann et al., 2020), there is little sign yet that the major toxic paradigms of the human world—economic growth, global markets and unrestrained affluence—will ever cease their dominance or their destructive impact on humans and the environment. The problems we face in restoring human and environmental wellbeing cannot be solved by conventional means alone or within current societal paradigms (Pyle, 2003; Wiedmann et al., 2020). Any solution will require us to end our separation from—our non-participation in—the natural world, to recognise the importance of intuitive knowledge as well as rational thought (Hamilton, 2002), and to value self-actualisation over self-interest, wisdom over wealth (Murtaza, 2011).

Building a new relationship with nature will be a beginning. Our connection with nature gives us the means to flourish as individuals and communities, the motivation to act to safeguard all our futures, and—if our connection is deep, profound and lasting enough—the wisdom with which to do it.

References

Aerts, R., Honnay, O., & Van Nieuwenhuyse, A. (2018). Biodiversity and human health: Mechanisms and evidence of the positive health effects of diversity in nature and green spaces. *British Medical Bulletin, 127*, 5–22.

Anderson, C. L., Monroy, M., & Keltner, D. (2018). Awe in nature heals: Evidence from military veterans, at-risk youth, and college students. *Emotion, 18*(8), 1195–1202.

Barton, J., Bragg, R., Pretty, J., Roberts, J., & Wood, C. (2016). The wilderness expedition: An effective life course intervention to improve young people's well-being and connectedness to nature. *The Journal of Experimental Education, 39*(1), 59–72.

Basu, M., Hashimoto, S., & Dasgupta, R. (2021). The mediating role of place attachment between nature connectedness and human well-being: Perspectives from Japan. *Sustainability Science, 15*, 849–862.

Baxter, D. E., & Pelletier, L. G. (2018). Is nature relatedness a basic human psychological need? A critical examination of the extant literature. *Canadian Psychology, 60*, 21–34.

Beery, T., Jönsson, K., & Elmberg, J. (2015). From environmental connectedness to sustainable futures: Topophilia and human affiliation with nature. *Sustainability, 7*, 8837–8854.

Bragg, E. A. (1996). Towards ecological self: Deep ecology meets constructionist self-theory. *Journal of Environmental Psychology, 16*, 93–108.

Capaldi, C. A., Dopko, R. L., & Zelenski, J. M. (2014). The relationship between nature connectedness and happiness: A meta-analysis. *Frontiers in Psychology, 5*, 976.

Carson, R. (1956). *The sense of wonder*. Harper Collins.

Castelo, N., White, K., & Goode, M. R. (2021). Nature promotes self-transcendence and prosocial behavior. *Journal of Environmental Psychology, 76*, 101639.

Chan, K. M. A., Balvanera, P., Benessaiah, K., et al. (2016). Why protect nature? Rethinking values and the environment. *PNAS, 113*, 1462–1465.

Chawla, L. (2020). Childhood nature connection and constructive hope: A review of research on connecting with nature and coping with environmental loss. *People and Nature, 2*, 619–642.

Dasgupta, P. (2021). *The economics of biodiversity: The Dasgupta review*. HM Treasury.

Davis, N., & Gatersleben, B. (2013). Transcendent experiences in wild and manicured settings: The influence of the trait 'connectedness to nature'. *Ecopsychology, 5*(2), 92–102.

Hamilton, C. (2002). Dualism and sustainability. *Ecological Economics, 42*, 89–99.

Howell, A. J., Passmore, H.-A., & Buro, K. (2013). Meaning in nature: Meaning in life as a mediator of the relationship between nature connectedness and well-being. *Journal of Happiness Studies, 14*, 1681–1696.

Ives, C. D., Giusti, M., Fischer, J., et al. (2017). Human–nature connection: A multidisciplinary review. *Current Opinion in Environmental Sustainability, 26–27*, 106–113.

Kahn, P. H., Jr. (1997). Developmental psychology and the biophilia hypothesis: Children's affiliation with nature. *Developmental Review, 17*, 1–61.

Keltner, D., & Piff, P. K. (2020). Self-transcendent awe as a moral grounding of wisdom. *Psychological Inquiry, 31*, 160–163.

Lambert, L., Lomas, T., van de Weijer, M. P., Passmore, H. A., Joshanloo, M., Harter, J., Ishikawa, Y., Lai, A., & Diener, E. (2020). Towards a greater global understanding of wellbeing: A proposal for a more inclusive measure. *International Journal of Wellbeing, 10*(2), 1–18.

Mackay, C. M., & Schmitt, M. T. (2019). Do people who feel connected to nature do more to protect it? A meta-analysis. *Journal of Environmental Psychology, 65*, 101323.

Martin, J. L., Maris, V., & Simberloff, D. S. (2016). The need to respect nature and its limits challenges society and conservation science. *PNAS, 113*(22), 6105–6112.

Martin, L., White, M. P., Hunt, A., Richardson, M., Pahl, S., & Burt, J. (2020). Nature contact, nature connectedness and associations with health, wellbeing and pro-environmental behaviours. *Journal of Environmental Psychology, 68*, 101389.

Mead, J., Fisher, Z., & Kemp, A. H. (2021). Moving beyond disciplinary silos towards a transdisciplinary model of wellbeing: An invited review. *Frontiers in Psychology, 12*, 1–10.

Mental Health Foundation. (2021). *How connecting with nature benefits our mental health*. Mental Health Foundation.

McEwan, K., Richardson, M., Sheffield, D., Ferguson, F. J., & Brindley, P. (2019). A smartphone app for improving mental health through connecting with urban nature. *International Journal of Environmental Research and Public Health, 16*(18), 3373.

Miller, J. R. (2005). Biodiversity conservation and the extinction of experience. *Trends in Ecology and Evolution, 20*(8), 430–434.

Murtaza, N. (2011). Pursuing self-interest or self-actualisation? From capitalism to a steady-state, wisdom economy. *Ecological Economics, 70*, 577–584.

Nelson, R. (1993). Searching for the lost arrow. In S. Kellert & E. O. Wilson (Eds.), *The Biophilia Hypothesis* (pp. 201–228). Island Press.

Orr, D. W. (1993). Love it or lose it: The coming biophilia revolution. In S. Kellert & E. O. Wilson (Eds.), *The Biophilia Hypothesis* (pp. 415–440). Island Press.

Passmore, H.-A., & Holder, M. D. (2017). Noticing nature: Individual and social benefits of a two-week intervention. *Journal of Positive Psychology, 12*(6), 537–546.

Passmore, H.-A., Martin, L., Richardson, M., White, M., Hunt, A., & Pahl, S. (2021). Parental/guardians' connection to nature better predicts children's nature connectedness than visits or area-level characteristics. *Ecopsychology, 13*(2), 103–113.

Pritchard, A., Richardson, M., Sheffield, D., & McEwan, K. (2019). The relationship between nature connectedness and eudaimonic well-being: A meta-analysis. *Journal of Happiness Studies, 21*(3), 1145–1167.

Pyle, R. M. (2003). Nature matrix: Reconnecting people and nature. *Oryx, 37*(2), 206–214.

Rabinowitz, P. M., Pappaioanou, M., Bardosh, K. L., & Conti, L. (2018). A planetary vision for one health. *BMJ Global Health, 3*(5), e001137.

Richardson, M., Dobson, J., Abson, D. J., Lumber, R., Hunt, A., Young, R., & Moorhouse, B. (2020). Applying the pathways to nature connectedness at a societal scale: A leverage points perspective. *Ecosystems and People, 16*(1), 387–401.

Richardson, M., Passmore, H. A., Barbett, L., Lumber, R., Thomas, R., & Hunt, A. (2020). The green care code: How nature connectedness and simple activities help explain pro-nature conservation behaviours. *People and Nature, 2*, 821–839.

Richardson, M., & Sheffield, D. (2017). Three good things in nature: Noticing nearby nature brings sustained increases in connection with nature. *Psyecology, 8*, 14.

Richardson, M., Passmore, H. A., Lumber, R., Thomas, R., & Hunt, A. (2021). Moments, not minutes: The nature-wellbeing relationship. *International Journal of Wellbeing, 11*(1), 8.

Riechers, M., Pătru-Dușe, I. A., & Balázsi, A. (2021). Leverage points to foster human–nature connectedness in cultural landscapes. *Ambio, 50*, 1670–1680.

Ripple, W. J., Wolf, C., Newsome, T. M., Barnard, P., & Moomaw, W. R. (2020). World scientists' warning of a climate emergency. *Bioscience, 70*, 8–12.

Shiota, M. N., Keltner, D., & Mossman, A. (2007). The nature of awe: Elicitors, appraisals, and effects on self-concept. *Cognition and Emotion, 21*, 944–963.

Soga, M., & Gaston, K. J. (2006). Extinction of experience: The loss of human–nature interactions. *Frontiers in Ecology and the Environment, 14*(2), 94–101.

Tree, I. (2018). *Wilding: The return of nature to a British farm*. Picador.

Trigwell, J. L., Francis, A. J. P., & Bagot, K. L. (2014). Nature connectedness and eudaimonic well-being: Spirituality as a potential mediator. *Ecopsychology, 6*, 1–11.

White, M., Elliott, L., Grellier, J., Economou, T., Bell, S., Bratman, G., Cirach, M., Gascon, M., Lima, M., Lohmus, M., et al. (2021). Associations between green/blue spaces and mental health across 18 countries. *Scientific Reports, 11*, 8903.

Wiedmann, T., Lenzen, M., Keyßer, L. T., & Steinberger, J. K. (2020). Scientists' warning on affluence. *Nature Communications, 11*, 3107.

Windhorst, E., & Williams, A. (2015). Growing up, naturally: The mental health legacy of early nature affiliation. *Ecopsychology, 7*, 115–125.

World Health Organization and Secretariat of the Convention on Biological Diversity. (2015). *Connecting global priorities: Biodiversity and human health.* WHO Press.

Wyles, K., White, M. P., Hattam, C., Pahl, S., & Austin, M. (2019). Nature connectedness and well-being from recent nature visits: The role of environment type and quality. *Environment and Behavior, 51*(2), 111–143.

Yang, Y., Hu, J., Fengjie, J. F., & Nguyen, B. (2018). From awe to ecological behavior: The mediating role of connectedness to nature. *Sustainability, 10*, 2477.

Zylstra, J. M., Knight, T. A., Esler, K. J., & Le Grange, L. L. L. (2014). Connectedness as a core conservation concern: An interdisciplinary review of theory and a call for practice. *Springer Science Reviews, 2*, 119–143.

CHAPTER 7

A Modern Framework of Wellbeing from the Perspective of Positive Environments

Víctor Corral-Verdugo, Martha Frias-Armenta, Nadia S. Corral-Frías, and Marc Yancy Lucas

Abstract Positive environments (PEs) are traditionally understood as contexts that provide the material and social resources necessary to allow individuals to flourish and experience physical and mental wellbeing. This traditional approach frames positive environments as primarily resource deliverers, which guarantee wellbeing by satisfying individual needs. However, recent conceptualizations consider PEs as more than resource providers, highlighting PEs as *sustainable environments* that require

V. Corral-Verdugo (✉) • N. S. Corral-Frías • M. Y. Lucas
Department of Psychology, University of Sonora, Hermosillo, Mexico
e-mail: victorcorral@sociales.uson.mx; nadia.corral@unison.mx; mylucas@email.arizona.edu

M. Frias-Armenta
Department of Law, University of Sonora, Hermosillo, Mexico
e-mail: martha.frias@unison.mx

A. H. Kemp, D. J. Edwards (eds.), *Broadening the Scope of Wellbeing Science*, https://doi.org/10.1007/978-3-031-18329-4_7

additional elements to maintain positivity and achieve individual wellbeing. These additional elements include: (1) enablers of environmentally responsible behaviors and (2) environmental challenges that must be overcome. From this perspective, PEs rely upon significant relationships between resources, enablers, and challenges. A balance between these three elements creates the opportunity for individual wellbeing, driven by environmental challenges, adequate access to the resources and the presence of enablers instigating responsible behaviors. This chapter examines theory and evidence demonstrating the association between wellbeing and positive environments, the ways in which the elements of PEs combine and balance available sociophysical resources, instigators of responsible behaviors, and environmental challenges.

Keywords Individual wellbeing • Positive environments • Resources • Enablers • Challenges • Responsible behaviors

Introduction

Understanding of wellbeing and positive psychological states are based on personal as well as environmental elements. Although psychology tends to focus on stress and the importance of personal (i.e., genetic, trait, behavioral) causes (Hobfoll, 2004), recent developments in the study of wellbeing have focused on the significant influences of the environment on positive psychological states (Ballew & Omoto, 2018). In some cases, those influences are analyzed from the perspective of *environmental qualities*, aspects or features of the milieu contributing to wellbeing and satisfaction of human needs (Van Kamp et al., 2003; Wu et al., 2013). According to this concept, individual exposure to environmental qualities provides psychological benefits. Environmental stability (Brumbach et al., 2009), safety (Vogel, 2012), naturalness (Luck et al., 2011), coherence (Yannick, 2007), and complexity (Roberts, 2007) are examples of environmental qualities that result in human wellbeing. Another perspective examines *affordances*: environmental properties that induce effective responses in organisms. Gibson established that "affordances of the environment are what it offers the animal, what it provides or furnishes, either for good or ill" (Gibson, 1979, p. 127). One additional approach establishes that the environment, as provider of resources (both physical and social) can meet individual needs and, thus, is a source of wellbeing (La Placa et al., 2013; Summers et al., 2012).

The common tenet of these utilitarian approaches (the environment as a provider of resources that guarantee wellbeing), if taken literally, can fail to consider a fundamental aspect of environmental influences on wellbeing: *the environment must be preserved to provide those necessary resources in the future and to sustain wellbeing and environmental quality.* One additional aspect that is often neglected in the utilitarian view of environmental positivity is the fact that the environment not only affords benefits but also offers challenges, and these challenges can constitute either a source of risk or opportunity for individuals to thrive (especially if individuals are willing to address these challenges). Gibson's (1979) definition of affordance establishes that the environment offers opportunities for good or ill. Paradoxically, even the "negative" aspects of affordances (i.e., challenges, risks) may be necessary for wellbeing (Wong et al., 2021). These challenges drive human adaptive responses to the environment and enable behaviors and psychological states conducive to wellbeing. Therefore, environmental resources, instigators of environmental conservation, and environmental challenges are fundamental elements of human positive scenarios, which we hereinafter name *positive environments*: "contexts that promote individual and collective benefits and that also influence human predispositions to conserve—in the long run—the sociophysical structures on which life depends" (Corral-Verdugo & Frías-Armenta, 2016, p. 965). These include natural and man-made environments.

This chapter examines theory and evidence demonstrating the association between wellbeing and positive environments (PEs), the ways in which the elements of PEs combine available sociophysical resources, enablers of responsible (prosocial and pro-environmental) behaviors, and environmental challenges. The theory and evidence suggest that the conjunction of these elements and their balance is one important base for the development and maintenance of wellbeing among individuals.

Wellbeing

Wellbeing, as a complex psychological construct, has been defined in various manners. Previous literature has suggested that is not simply the absence of illbeing, given that it is possible to experience wellbeing during ill health, and that negative affect may lead to growth and subsequent positive outcomes (Kashdan & Biswas-Diener, 2015). Research has likewise begun to move away from definitions that equate wellbeing with quality of life, happiness, or economic wealth. More recently, wellbeing has been defined as psychological experience, promoted by connection to self, community and environment, supported by healthy vagal function, impacted on by sociocontextual factors that lie beyond the control of the individual (Mead et al., 2021, p. 1; see also Corral, Pato, & Torres, 2021; Kemp & Fisher, 2022). Recent work shows how these frameworks have begun to be applied within the healthcare and education sectors (Kemp & Fisher, 2022). Recent calls for transdisciplinary approaches for the study of wellbeing emphasize a need to place the person within their community, environment, and sociocultural contexts (Kemp & Fisher, 2022; Mead et al., 2021). This means that wellbeing should be assessed in context, considering its interdependence and inter-relationships with multiple domains and levels of scale.

Positive Environments

The field of environmental psychology (EP) examines human behavior and wellbeing as it relates to the sociophysical environment (Stokols & Altman, 1987). More specifically, EP explores the environmental context of human behavior and wellbeing (De Young, 2013), acknowledging the mutual influence ofindividuals with their environments (Bell & Sundstrom, 1997). This EP definition implies that the environment affects wellbeing, but wellbeing also influences environmental quality through, for example, the facilitation of pro-environmental behaviors.

As previously outlined, a positive environment provides resources that enable the individual to successfully face and overcome environmental challenges, while simultaneously requiring the conservation of those resources for future needs (Corral-Verdugo et al., 2021). The PE resulting from the combination of these three elements influences (and is influenced by) the sustainable behaviors of individuals and, likewise of equal importance, personal wellbeing. These reciprocal influences mean that, in a positive environment, individuals engage in sustainable behaviors and experience wellbeing, which produce a recursive chain of positive interdependencies between settings, behaviors, and wellbeing (Corral-Verdugo & Frías-Armenta, 2016). Figure 7.1 is a representation of a PE, its elements, and interrelations.

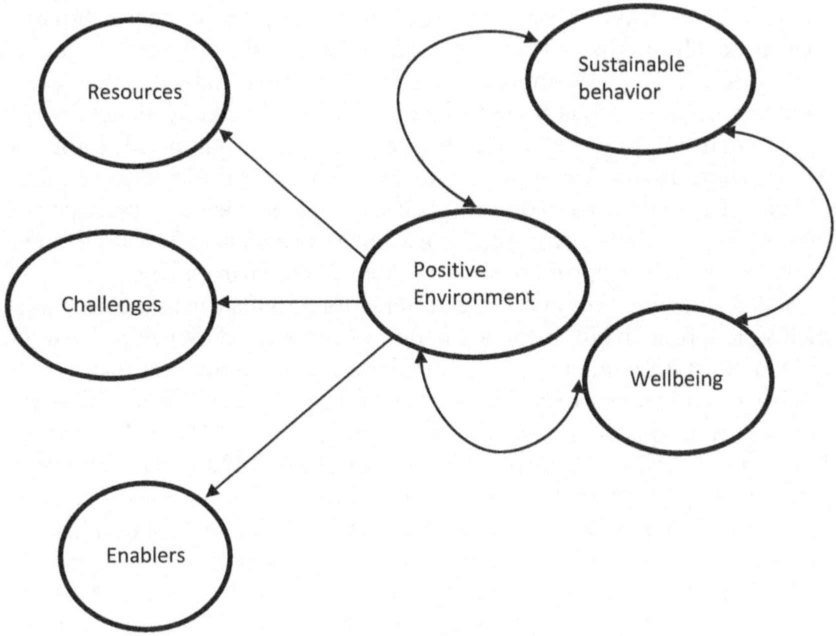

Fig. 7.1 Schematic representation of a positive environment, its elements, and interrelations

Elements of a Positive Environment

Environmental Resources

The sociophysical environment provides affordances for satisfying needs and fostering positive psychological states. It can be understood as an amalgamation of cultural/contextual norms, expectations, beliefs, and the material aspects it encompasses. The resources it provides may be tangible (physical) or immaterial (social) and, examined in concert, represent a holistic portrait of potential mechanisms available to the individual in the pursuit of wellbeing. The physical environment may provide sustenance, shelter, mobility, and settings for fitness and psychological restoration (Leviston et al., 2018), while social resources embedded in familial, professional, and community networks can support collective positivity, voluntary work, and shared social capital (Diener, 2012). Infrastructure such as that related to healthcare provision, labor, transportation, and education can further contribute to available sociophysical environmental affordance (Selian & McKnight, 2017). The relative availability of such resources facilitates wellbeing (Pilkington et al., 2012) and may influence the degree to which the sociophysical environment and its related affordances create a sentiment of physical, psychological, and social safety (Aldridge & McChesney, 2018). A comprehensive inventory of the elements required for an environment to promote positivity and wellbeing is perhaps too prescriptive to fully encompass the human experience; however, some underlying factors appear consistent among such environments.

Social elements that encourage interaction, conflict management, and inclusion appear to aid in the formation of positive relationships between individuals. Likewise, perceptions of institutional and social fairness, equal treatment from roles of authority, and affirmative treatment within the social setting appear to foster positive environments (Frias-Armenta & Corral-Frias, 2021). Similarly, elements of physical space drive and reinforce behavior settings that can be tailored to promote wellbeing and foster positivity. The scope can range from small-scale phenomenon like the interior design and spatial orientation of a room or building to macro-scale, comprehensive planning initiatives. Generally, the types of elements that promote positivity and wellbeing are characterized by inclusion, physical comfort and safety, and affirmative relationships between individuals that provide for necessities and afford the tools to meet challenges presented by the environment.

Wellbeing-Resources Relationship

Availability of environmental resources represents a reciprocal relationship with human wellbeing, which implies that in a positive environment, a double benefit is obtained. For example, ecosystems provide benefits to people, and once their basic needs are met, they may work to protect the integrity of those ecosystems for the future (Leviston et al., 2018). Evidence suggests social resources represent a reciprocal relationship with wellbeing, which itself results from availability of social support. This type of social safety net may motivate individuals to pay forward the support they received, further sustaining wellbeing and ensuring future resources (Pilkington et al., 2012). The emergence of a dual benefit in the relationship between resource availability and human wellbeing begs an important question: what factors determine this type of positive reciprocity? Put another way, under what circumstances does human wellbeing lead to the preservation of natural and social resources (and vice versa)? Some clues are suggested by previous research examining the satisfaction of basic human survival needs; if these needs are met, the wellbeing resulting from use of natural resources can drive the pro-environmental behaviors of individuals (Leviston et al., 2018). Corral-Verdugo and Frías-Armenta (2016) suggest that environmental positivity provided by sociophysical resources and pro-environmental instigators produces wellbeing which, in turn, increases environmental positivity. Since environmental resources are part of positive environments, this reciprocal relationship is implicit in the PE-wellbeing association (that is why such a relationship is not represented in Fig. 7.1).

Pro-environmental Enablers

The expected responsible actions of individuals in a positive environment are triggered by situations that promote pro-environmental and prosocial behaviors; these behaviors, in turn, guarantee the sustained conservation of that environment and, consequently, the possibility of obtaining further wellbeing. Enablers of responsible actions include the availability of pro-environmental affordances, those opportunities provided by the environment to act in a sustainable way. Accessible recycling systems, safe urban bikeways, and accessibility to urban nature represent affordances instigating pro-environmental behaviors. Similarly, comfortable and efficient public transportation that provides a viable alternative to individual vehicle use, multiple land-use planning, affordable solar panels, and availability of environment-friendly products and services may encourage sustainable actions.

Information which provides the tools to cope with and overcome environmental barriers represents another important enabler. This information can be provided through environmental education programs, either formal or informal, or through community campaigns targeting the conservation of natural and social resources (Mustam & Daniel, 2018). Also, social models of responsible behavior serve as enablers of appropriate actions in a positive environment. Focus theory of normative conduct establishes that the sustainable behavior of individuals is influenced by the sociophysical milieu, the actions of others, and the behavior settings that drive conduct. In a pro-sustainable society, the expectation is to engage in daily prosocial and pro-environmental behaviors. Also, in such society, models of responsible behavior from authoritative figures are prominent (Abrahamse, 2019). For example, Flynn et al. (2021) found that front-line leaders of health services influenced staff effort to engage in an interventional community program. These leaders were "hands on" and used their influence to communicate the impact and success of the intervention.

Perceived legitimacy of authorities when addressing important social issues, and government programs guaranteeing access to social justice, and equity are another critical source of environmental enablers that foster sustainable behaviors (Corral-Verdugo et al., 2021). Studies have demonstrated that relevant information providing guidelines for responding to a social crisis (Lau et al., 2020) and perceived legitimacy of authority in managing that crisis (Hartley & Jarvis, 2020) are useful in encouraging public compliance with responsible behaviors, including those in the context of the recent COVID-19 pandemic (Corral-Verdugo, Corral-Frías, Frías-Armenta, Lucas, & Peña, 2021).

The presence of pro-environmental enablers instigates the practice of sustainable behaviors that can result in wellbeing and other positive psychological states. This suggests that the influence of enablers on wellbeing is indirect and mediated by sustainable behaviors. The mediating influence of enablers on wellbeing through SB is also implicit in Fig. 7.1 since enablers are part of a positive environment. It is likely that instigators of these responsible behaviors lead to wellbeing via feelings of satisfaction and self-efficacy produced by the feeling that an individual has engaged in prosocial and pro-environmental activities (De Young, 2013).

Environmental Challenges

Environmental challenges include poverty, inequity, disease, violence, climate change, and environmental degradation, among others. One of the most pervasive challenges for humanity is poverty, with 82% of the world population living with less than 20 dollars per day (Our World in Data, 2021). These individuals lack the indispensable resources to cover their basic needs, which can limit the development and opportunities available to children and lead to undernourishment, poor health, and limited access to social support (Human Development Report, 2014). Countries experiencing the highest levels of poverty likewise represent those with the greatest degree of socioeconomic and gender-based inequity. In many countries, women who care for the home receive little to no individual income while being responsible for the care of other family members, which can result in a lack of individual agency. Discrimination by race, ethnicity, and sexual orientation represent further challenges. Furthermore, public health crises like the COVID-19 pandemic can further exacerbate issues of inequity, poverty, and environmental issues like climate change.

Violence is usually represented as aggressive behavior against others; however, the economic hardship experienced by a significant portion of the world population represents the most common, and perhaps under-recognized, form of violence. In the same vein, institutional violence represents a mechanism of structural violence that frequently affects poor people. Each year around 1.6 million people lose their lives to violence (WHO, 2002).

The climate crisis is also a major overarching challenge. Ripple et al. (2021) identified a significant surge in climate-related disasters since 2019, including devastating flooding in South America and Southeast Asia; heat waves and wildfires in Australia and the Western United States; an extraordinary Atlantic hurricane season; and devastating cyclones in Africa, South Asia, and the West Pacific.

Environmental degradation, a huge challenge for humankind, is caused by the concentration of pollutants, which causes harmful changes in the biosphere, a reduction of biodiversity, and health problems (Arneth et al., 2020). Industrialized countries produce most of the world's pollution, and the most affected are the disadvantaged people. Each year it is estimated that 600,000 children die from respiratory problems caused by air pollution (WHO, 2018). All these problems can be taken as challenges to action in the search of environmental positivity and wellbeing.

Challenges present the opportunity to be overcome, allowing individuals agency to take control of their lives, make choices, and learn from the consequences of those choices. This suggests the potential for a positive outcome by taking risks when the individual can successfully navigate that risk (Mishra, Barclay, & Sparks, 2017). Accepting and identifying challenges and looking for ways to surpass them are necessary for converting them into opportunities and successful outcomes for the individual and societies (Jacobs, 2011).

The presence of challenges in one's environment can produce various outcomes: an individual can (1) experience challenge and not act, (2) face and cope with challenges but lack the ability to successfully overcome them, and (3) successfully overcome the challenge. Theoretically, outcome three represents a positive environment that can foster individual wellbeing. Self-efficacy, sense of self-competency in dealing with the challenge, is a likely triggering mechanism for wellbeing in this situation (De Young, 2013). The Positive Environment Model suggests a bidirectional relationship between environmental challenges and wellbeing. For example, individuals reporting a greater degree of wellbeing are more likely to identify challenges that, in turn, serve to maintain wellbeing. Moreover, as explained by the theories of Flow (Csikszentmihalyi, 2007) and Dynamic Equilibrium (Kloep et al., 2009) manageable challenges within the range of skills and ability of the individual are necessary to maintaining an adequate level of wellbeing.

Sustainable Behaviors

Sustainable behaviors (SBs) are actions aimed at protecting people and the natural milieu, both at present and for the future (Tapia et al., 2013). A particular classification suggests that sustainable behaviors comprise actions aimed at self-care, caring for others, and caring for the biosphere (Corral-Verdugo, Pato, & Torres, 2021). Self-care behaviors encompass activities intended to maintain one's own physical and mental health, such as exercising, consuming healthy food, practicing meditation, and fomenting loving relationships. Caring for others includes the practice of altruistic and equitable behaviors. Donating money for charities and the poor, assisting sick people, donating blood, and supporting people in need can be considered altruistic behaviors. Fairness in everyday activities, avoiding discrimination based on race or other personal characteristics, and providing opportunities for marginalized populations represent equitable behavior. Finally, caring for the biosphere involves pro-ecological and frugal behaviors. Pro-ecological actions involve caring for the integrity of resources such as water, air, energy, soil, and ecosystems; frugal behaviors are those that avoid wasting resources and induce minimal ecological impact. The Positive Environment Model establishes that SBs are stimulated by pro-environmental enablers but also by the presence of sociophysical resources and environmental challenges (Corral-Verdugo et al., 2021). This implies that, in a positive environment, individuals and groups tend to engage in prosocial and pro-environmental behavior that likewise contribute to environmental positivity.

Potentially counterintuitive, practicing sustainable behaviors often produces positive psychological outcomes for the individual: caring for self, others, and the environment makes people feel good. Evidence suggests individuals who routinely engage in healthy behaviors (i.e., healthy eating and exercising) report more happiness than those who do not (Veenhoven, 2019). Likewise, people who engage in prosocial activities including altruistic and equitable behaviors experience subjective and psychological well-being, as well as reporting feelings of satisfaction, self-efficacy, and pleasure (Moll et al., 2006). The practice of pro-environmental behavior is also intrinsically reinforced by at least some of these positive psychological states (Brown & Kasser, 2005). Why does this happen? Although no conclusive answer to this question exists, it can be speculated that those pleasurable, positive feelings were selected in the evolution of humankind, as reinforcers of behaviors that are favorable for the individual (self-care) or group (caring for others) and thus serve to promote and sustain a positive environment.

Wellbeing as Balance Between Elements of the Positive Environment

The *Dynamic Equilibrium Theory* (Kloep et al., 2009) holds that an excess of challenges with respect to resources causes an imbalance and vice versa. If challenges overburden available resources, to overcome them, a sustainable positive environment may be difficult to foster. Conversely, when resource availability makes challenges too easy to navigate, it can lead to boredom and conformity. By definition, neither scenario achieves a sustainable positive environment supportive of wellbeing.

The positive environment perspective further examines the role of pro-environmental enablers in this tripartite (resources, challenges, enablers) balancing act. A positive environment is sustainable and can support people over a long period of time if it manages a relative balance between these three basic elements. For example, if an environment only provides resources without asking for its conservation, sooner or later it will be depleted, and the satisfaction of needs will no longer be assured. Inversely, if resources are available but people are not able to face challenges, they will be unable to reach their highest degree of potential self-actualization. An environment that demands a continuous preservation of its resources without offering material and intangible satisfiers cannot provide wellbeing. Further, an environment of unsurmountable challenges that does not afford resources to overcome them makes the promotion of individual and social wellbeing difficult. The need for environmental preservation can be both a challenge and an instigator of environmentally responsible behaviors, which means that the enablers of sustainable actions often follow threats of environmental deterioration.

Future Directions

The model of environmental positivity has been tested although not in its entirety. One study assessed family positivity, indicated by availability of physical resources (housing habitability) and family functioning, which was found to be directly and significantly associated with personal wellbeing (Durón et al., 2019). In another, a positive school environment had a negative effect on adolescent antisocial behavior, via mood and anxiety disorders as well as in interaction with agreeableness (Frías-Armenta & Corral-Frías, 2021). Although promising, these studies indicate that further steps should be taken to model a PE. For instance, the

PE model establishes that, more than the ***individual contribution*** of each element to positivity, the ***interrelation*** and ***balance*** between those elements are key aspects to achieve high levels of wellbeing. These assumptions could be empirically tested in diverse ways. Using multivariate analyses, the interrelation between resources, enablers, challenges, and wellbeing could be assessed and tested. One potential method could include model 1, testing covariances between each separate element (resources, enablers, challenges) of a PE and wellbeing (see Fig. 7.2); model 2, combining two elements to build a PE factor and testing its association with wellbeing (see Fig. 7.3), and finally, model 3, combining the three elements to build the PE factor and testing its covariance with wellbeing (see Fig. 7.4).

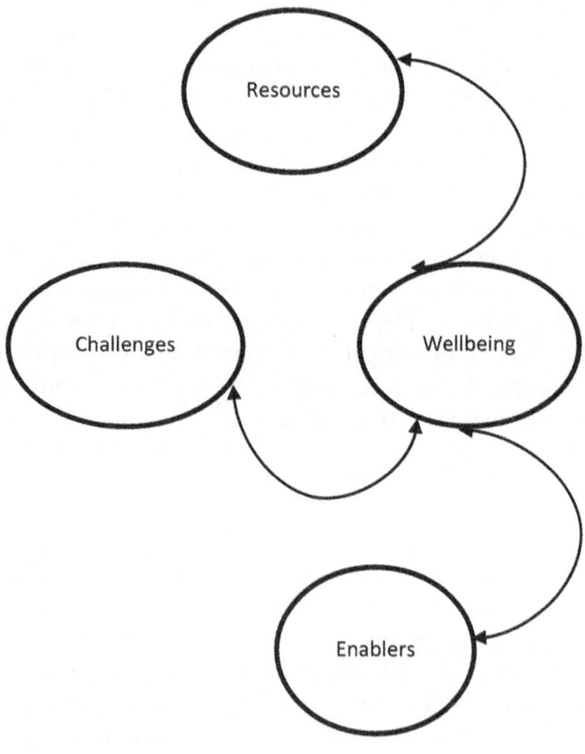

Fig. 7.2 Representation of a model testing covariances between wellbeing and resources, challenges and enablers

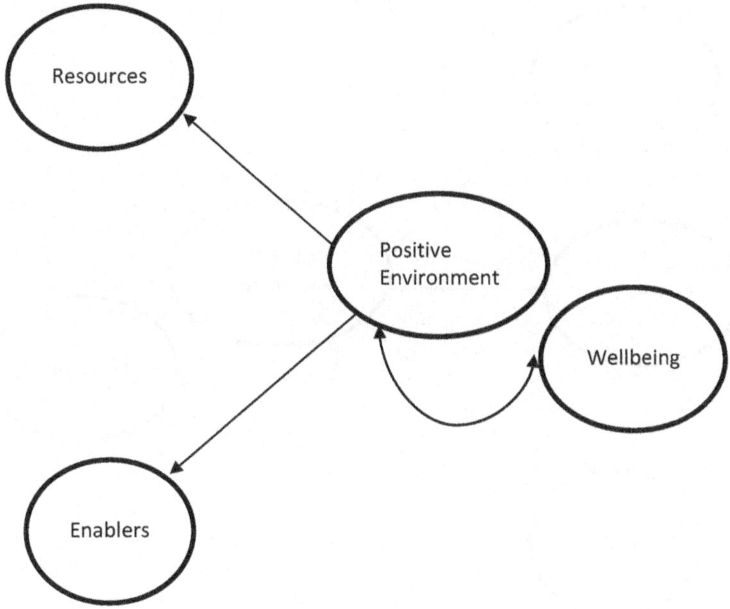

Fig. 7.3 Combining two elements (resources, enablers) to build a positive environment factor and testing its association with wellbeing

The ultimate test would imply comparing the strength of association between wellbeing and PE throughout all these models. The strongest association is expected to occur in the third model. In addition, balance between PE elements could be assessed in the context of factor analysis. Similar scores and contribution (i.e., *lambdas*) from the three elements would indicate a balanced contribution of resources, challenges, and enablers to environmental positivity. The internal consistency of the PE factor should be a supplementary indicator of balance, demonstrating that the elements are not only interrelated but also balanced.

A recent study of environmental and psychological correlates of precautionary measures against COVID-19 (Corral-Verdugo, Corral-Frías, Frías-Armenta, Lucas, & Peña, 2021) seemed to confirm the assumptions of interrelationship and balance, which suggested infrastructural, family,

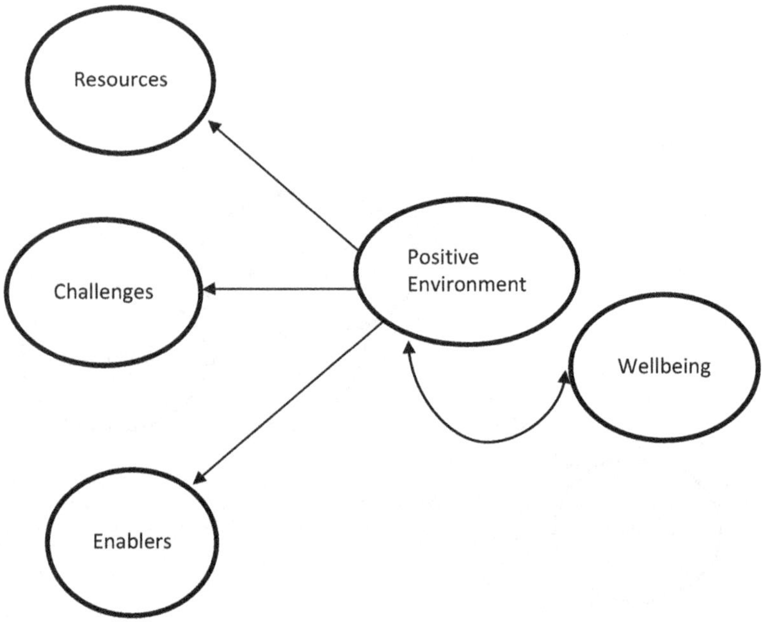

Fig. 7.4 Combining resources, challenges and enablers to build a positive environment factor and testing its association with wellbeing

and social resources, along with enablers (information, legitimacy of authorities) of responsible behaviors and environmental challenges (threat of contagion, social pressure) constituted a PE factor, which highly and significantly covaried with both responsible behaviors (self-care, precautionary measures against COVID-19), and hedonic wellbeing. However, the study did not compare restricted models (1, 2) against the whole PE model (3), so that this test is expected in further studies.

In any case, extant evidence indicates that a positive environment promotes wellbeing and that wellbeing will also facilitate levels of environmental positivity. Given these results, efforts to deepen this relationship are to be expected and will provide useful information for the further design and development of positive environments.

References

Abrahamse, W. (2019). *Encouraging pro-environmental behaviour: What works, what doesn't, and why*. Academic Press.

Aldridge, J. M., & McChesney, K. (2018). The relationships between school climate and adolescent mental health and wellbeing: A systematic literature review. *International Journal of Educational Research, 88*, 121–145.

Arneth, A., Shin, Y. J., Leadley, P., Rondinini, C., Bukvareva, E., Kolb, M., et al. (2020). Post-2020 biodiversity targets need to embrace climate change. *Proceedings of the National Academy of Sciences, 117*(49), 30882–30891.

Ballew, M. T., & Omoto, A. M. (2018). Absorption: How nature experiences promote awe and other positive emotions. *Ecopsychology, 10*(1), 26–35.

Bell, P. A., & Sundstrom, E. D. (1997). Environmental psychology: evolution and current status. 1997). In D. Halpern & A. Voiskounsky (Eds.), *States of mind: American and post-Soviet perspectives on contemporary issues in psychology* (pp. 373–390). Oxford, UK: Oxford University Press.

Brown, K. W., & Kasser, T. (2005). Are psychological and ecological well-being compatible? The role of values, mindfulness, and lifestyle. *Social Indicators Research, 74*, 349–368.

Brumbach, B. H., Figueredo, A. J., & Ellis, B. J. (2009). Effects of harsh and unpredictable environments in adolescence on development of life history strategies. *Human Nature, 20*, 25–51.

Corral-Verdugo, V., & Frías-Armenta, M. (2016). The sustainability of positive environments. *Environment, Development, and Sustainability, 18*, 965–984.

Corral-Verdugo, V., Pato, C., & Torres, N. (2021). Testing a tridimensional model of sustainable behavior: Self-care, caring for others and caring for the planet. *Environment, Development, and Sustainability. 23*, 12867–12882.

Corral-Verdugo, V., Corral-Frías, NS., Frías-Armenta, M., Lucas, M. Y., & Peña, E. (2021). Positive environments and precautionary behaviors during the COVID-19 outbreak. *Frontiers in Psychology, 12*, 624155. https://doi.org/10.3389/fpsyg.2021.624155

Csikszentmihalyi, M. (2007). *Finding flow*. Basic Books.

De Young, R. (2013). Environmental psychology overview. In A. H. Huffman & S. R. Klein (Eds.), *Green Organizations: Driving Change with IO Psychology* (pp. 17–33). Routledge.

Diener, E. (2012). New findings and future directions for subjective well-being research. *American Psychologist, 67*(8), 590–597.

Durón, F., Tapia, C., Corral, V., & Fraijo, B. (2019). Ambiente familiar positivo y bienestar personal: Comparación entre población urbana y rural. *Revista Costarricense de Psicología, 38*(2), 225–239.

Flynn, R., Mrklas, K., Campbell, A., Wasylak, T., & Scott, S. D. (2021). Contextual factors and mechanisms that influence sustainability: A realist evaluation of two scaled, multi-component interventions. *BMC Health Services Research, 21*(1), 1–17.

Frías-Armenta, M., & Corral-Frías, M. (2021). Positive university environment and agreeableness as protective factors against antisocial behavior in Mexican university students. *Frontiers in Psychology, 12*, 662146.

Gibson, J. J. (1979). *The ecological approach to visual perception*. Houghton Mifflin Harcourt (HMH).

Hartley, K., & Jarvis, D. S. (2020). Policymaking in a low-trust state: Legitimacy, state capacity, and responses to COVID-19 in Hong Kong. *Policy and Soc, 39*, 403–423. https://doi.org/10.1080/14494035.2020.1783791

Hobfoll, S. E. (2004). *Stress, culture, and community: The psychology and philosophy of stress*. Springer Science & Business Media.

Human Development Report (2014). Retrieved from: https://www1.undp.org/content/undp/es/home/librarypage/hdr/2014-human-development-report.html

Jacobs, G. (2011). The turn towards unity: Converting crises into opportunities. *Cadmus, 1*(3), 106–120.

Kashdan, T., & Biswas-Diener, R. (2015). *The upside of your dark side: Why being your whole self–not just your "Good" self–drives success and fulfillment*. Plume Books.

Kemp, A. H., & Fisher, Z. (2022). Wellbeing, whole health and societal transformation: Theoretical insights and practical applications. *Global Advances in Health and Medicine, 11*, 21649561211073077. https://doi.org/10.1177/21649561211073077

Kloep, M., Hendry, L., & Saunders, D. (2009). A new perspective on human development. *Conference of the International Journal of Arts and Sciences, 1*, 332–343.

Lau, H., Khosrawipour, V., Kocbach, P., Mikolajczyk, A., Ichii, H., Schubert, J., & Khosrawipour, T. (2020). Internationally lost COVID-19 cases. *Journal of Microbiology, Immunology, and Infection, 53*, 454–458. https://doi.org/10.1016/j.jmii.2020.03.013

La Placa, V., McNaught, A., & Knight, A. (2013). Discourse on wellbeing in research and practice. *International Journal of Wellbeing, 3*(1), 116–125.

Leviston, Z., Walker, I., Green, M., & Price, J. (2018). Linkages between ecosystem services and human wellbeing: A nexus webs approach. *Ecological Indicators, 93*, 658–668.

Luck, G., Davidson, P., Boxall, D., & y Smallbone, L. (2011). Relations between urban bird and plant communities and human well-being and connection to nature. *Conservation Biology, 25*, 816–826.

Mead, J., Fisher, Z., & Kemp, A. (2021). Moving beyond disciplinary silos towards a transdisciplinary model of wellbeing: An invited review. *Frontiers in Psychology, 12*, 642093.

Mishra, S., Barclay, P., & Sparks, A. (2017). The relative state model: Integrating need-based and ability-based pathways to risk-taking. *Personality and Social Psychology Review, 21*(2), 176–198.

Moll, J., Krueger, F., Zahn, R., Pardini, M., de Oliveira-Souza, R., & Grafman, J. (2006). Human fronto–mesolimbic networks guide decisions about charitable donation. *Proceedings of the National Academy of Sciences, 103*(42), 15623–15628.

Mustam, B., & Daniel, E. S. (2018). Informal and formal environmental education infusion: Actions of Malaysian teachers and parents among students in a polluted area. *Malaysian Online Journal of Educational Sciences, 4*(1), 9–20.

Our World in Data (2021). Retrieved from: https://ourworldindata.org/higher-poverty-global-line.

Pilkington, P. D., Windsor, T. D., & Crisp, D. A. (2012). Volunteering and subjective well-being in midlife and older adults: The role of supportive social networks. *Journals of Gerontology Series B: Psychological Sciences and Social Sciences, 67*(2), 249–260.

Ripple, W. J., Wolf, C., Newsome, T. M., Gregg, J. W., Lenton, T. M., Palomo, I., et al. (2021). World scientists' warning of a climate emergency 2021. *Bioscience, 71*(9), 894–898.

Roberts, M. N. (2007). *Complexity and aesthetic preference for diverse visual stimuli*. Doctoral dissertation. Department of Psychology, Universitat de les Illes Balears.

Selian, A. N., & McKnight, L. (2017). The role of technology in the history of well-being: Transformative market phenomena over time. In R. Estes & M. Sirgy (Eds.), *The Pursuit of Human Well-being. International Handbooks of Quality-of-Life* (pp. 639–687). Springer. https://doi.org/10.1007/978-3-319-39101-4_19

Stokols, D., & Altman, I. (1987). Introduction. In D. Stokols & I. Altman (Eds.), *Handbook of Environmental Psychology* (pp. 1–4). Wiley & Sons.

Summers, J. K., Smith, L. M., Case, J. L., & Linthurst, R. A. (2012). A review of the elements of human well-being with an emphasis on the contribution of ecosystem services. *Ambio, 41*(4), 327–340.

Tapia, C., Corral, V., Fraijo, B., & Durón, F. (2013). Assessing sustainable behavior and its correlates: a measure of pro-ecological, frugal, altruistic and equitable actions. *Sustainability, 5*, 711–723.

Van Kamp, I., Leidelmeijer, K., Marsman, G., & De Hollander, A. (2003). Urban environmental quality and human well-being: Towards a conceptual framework and demarcation of concepts; a literature study. *Landscape and Urban Planning, 65*(1–2), 5–18.

Veenhoven, R. (2019). Will healthy eating make you happier? A research synthesis using an online findings archive. *Applied Research Quality Life, 16*, 221–240. https://doi.org/10.1007/s11482-019-09748-7

Vogel, D. (2012). *The politics of precaution: Regulating health, safety, and environmental risks in Europe and the United States.* Princeton University Press.
WHO. (2002). *World report on violence and health.* WHO.
WHO (2018). Retrieved from: https://www.who.int/news/item/29-10-2018-more-than-90-of-the-world%e2%80%99s-children-breathe-toxic-air-every-day.
Wong, P. T., Mayer, C. H., & Arslan, G. (2021). COVID-19 and existential positive psychology (PP2. 0): The new science of self-transcendence. *Frontiers in Psychology, 12*, 800308.
Wu, D. W. L., DiGiacomo, A., & Kingstone, A. (2013). A sustainable building promotes pro-environmental behavior: An observational study on food disposal. *PLoS One, 8*(1), e53856.
Yannick, J. (2007). Architectural lessons from environmental psychology: The case of biophilic architecture. *General Review of Psychology, 11*, 305–328.

CHAPTER 8

A Public Health Perspective on Wellbeing

Nerys Edmonds

Abstract This chapter places the promotion of wellbeing in the context of public health practice and explores enduring challenges, practical solutions, and synergies with the field of positive psychology and contextual behavioural science. Public health applies multilevel frameworks of health and wellbeing to inform and develop policies and programmes to improve population health and wellbeing outcomes, distinct from individual outcomes. How the concept of wellbeing is translated into policy and practice matters, particularly in the face of recent and emerging political, societal, and public health challenges in the UK such as "Brexit", the COVID-19 pandemic, and the climate emergency—where the need to protect and promote the population's health and wellbeing and address inequalities have become ever more pressing. Specifically, the chapter provides examples of how public health tools such as Health Impact Assessment (HIA) and Mental Wellbeing Impact Assessment (MWIA) can facilitate upstream multi-sectoral collaboration to integrate wellbeing into policy and programmes.

N. Edmonds (✉)
Public Health Wales, Cardiff, Wales, UK
e-mail: Nerys.S.Edmonds@wales.nhs.uk

A. H. Kemp, D. J. Edwards (eds.), *Broadening the Scope of Wellbeing Science*, https://doi.org/10.1007/978-3-031-18329-4_8

Keywords Population health • Health promotion • Impact assessment • Wellbeing • Mental wellbeing • Health equity

Abbreviations

ACT	Acceptance and Commitment Therapy
HIA	Health Impact Assessment
HiAP	Health in All Policies
MWIA	Mental Wellbeing Impact Assessment
QALY	Quality-Adjusted Life Year
WALY	Wellbeing-Adjusted Life Year
WELLBY	Wellbeing Year
WHO	World Health Organization

Introduction

This chapter places the promotion of wellbeing in the context of public health and health promotion practice. Public health and health promotion apply multilevel frameworks of health and wellbeing to inform and develop policies and programmes to improve population health and wellbeing outcomes, which are distinct from individual outcomes. Public health practice engages in action across systems to influence the socioeconomic and environmental factors and contexts that affect wellbeing outcomes, which may be relevant to the broadening scope (Lomas et al., 2020) and transdisciplinary thinking (Mead et al., 2021) of wellbeing science. This chapter provides examples of how public health tools such as Health Impact Assessment (HIA) and Mental Wellbeing Impact Assessment (MWIA) can facilitate upstream multi-sectoral collaboration to integrate wellbeing into policy and programmes.

About Public Health and Health Promotion

Public health is "*the science and art of preventing disease, prolonging life and promoting health through the organized efforts of society*" (Wanless, 2003). Increasingly, the term "population health" is being used to distinguish multi-sectoral action aimed at improving the health and wellbeing of an entire population from the professional discipline of public health

(Buck et al., 2018). Improving population health and wellbeing requires action to address enduring socioeconomic inequities and engagement with actors and organisations outside of the health sector (Marmot et al., 2010) (see Box 8.1). The terms public health and population health are used interchangeably in this chapter.

Box 8.1 Inequities in Health and Wellbeing in Wales

- The gap in life expectancy between the most and least deprived areas of Wales is 9.0 years for males and 7.4 years for females (2017–19).
- The gap in healthy life expectancy between the most and least deprived areas of Wales is 17.0 years for males and 18.3 years for females (2017–19).
- 88% of persons over 16 living in the least deprived areas of Wales reported high life satisfaction compared to 74% in the most deprived areas (2016/17).
- 86% of persons over 16 in employment in Wales reported high life satisfaction compared to 66% of those adults who were unemployed (2016/17).
- 9% of persons over 16 living in the least deprived areas of Wales reported low mental wellbeing compared to 17% in the most deprived areas (2016/17).

Sources: Public Health Wales Observatory (2020) and Welsh Government (2021a)

In common with positive psychology (Lomas et al., 2020), the evolution of public health has been described in "waves", often coinciding with major changes in society and the collective interpretation of "health" (Hanlon et al., 2011). A proposed "fifth wave" of public health was put forward by Hanlon et al. (2011) in response to complex population health challenges such as obesity, increasing mental ill-health, rising inequalities, and unsustainable energy use. Features of this "fifth wave" include recognising and influencing complex adaptive systems, greater collaboration and sharing of learning, enabling positive change rather than "fixing" problems and people, and more focus on the subjective elements of health and wellbeing (Hanlon et al., 2011). These waves are also consistent with other wellbeing approaches such as contextual behavioural science and acceptance and commitment therapy, which emphasise positive change and a commitment to values-based living rather than the direct amelioration of "maladaptive" thinking (Hayes, Strosahl, & Wilson, 2011; Edwards et al., 2019).

The World Health Organization (WHO) describes 10 domains of essential public health provision (WHO, 2022). The main focus of this chapter is with the "health promotion" domain, which was defined at the first international conference on health promotion in Ottawa in 1986 (WHO, 2009) as:

> *the process of enabling people to increase control over, and to improve, their health. To reach a state of complete physical, mental and social* ***wellbeing****, an individual or group must be able to identify and to realize aspirations, to satisfy needs, and to change or cope with the environment.... health promotion is not just the responsibility of the health sector, but goes beyond healthy life-styles to* ***wellbeing*** (WHO, 2009, p. 1)

The definition of health promotion and conceptual framework set out by the Ottawa conference are important here for several reasons. Firstly, a holistic definition of wellbeing has been central to health promotion from its inception. Secondly, health promotion action (including promoting wellbeing) is defined as operating at multiple levels (see Box 8.2) and across the "settings" where people live, work, and learn and so has relevance to current debates in the field of wellbeing around broadening the scope of positive psychology to include the contexts in which wellbeing is enhanced or diminished, and the systemic factors which influence this (Kemp & Fisher, 2022; Kern et al., 2019; Lomas et al., 2020).

Box 8.2 Levels of Health Promotion Action in the Ottawa Charter (WHO, 2009)

- *Individual level*: Promoting personal skills for health and wellbeing
- *Community level*: Strengthening community action
- *Organisational level*: Creating supportive living environments, re-orientating health services to a prevention focus
- *System wide/policy levels*: Building healthy public policy

Wellbeing in Public Health and Health Promotion

Public health is strongly aligned to evidence-based medicine (Berghs et al., 2019). Measures such as "Quality- Adjusted Life Year" (QALY) are central to the quantification of health gains in cost-effectiveness analysis that inform the commissioning of health services and preventative interventions (National Institute for Health and Care Excellence, 2012). Therefore, work to establish the validity and utility of wellbeing measures at population level to inform public health practice has been an important, and contested, development (Bartlett & Coles, 1998a, 1998b; de Cates et al., 2015; Mehta et al., 2015; Stewart-Brown et al., 2015). In this context, progress towards a "QALY" type measure that integrates wellbeing such as the "WALY" (Johnson et al., 2016) or "WELLBY" (Layard & Oparina, 2021) are significant.

Mental health and subjective wellbeing were neglected in public health until the 1990s when an agenda that focused on the wider social and economic relevance of mental health as a major contributor to the "burden of disease" and population health developed beyond a historic focus on treatment of mental illness (Wahlbeck, 2015; Walker et al., 2019). Governments have begun to "*realise the value of a mentally healthy population*" (Senior et al., 2020) and this has led to a greater focus on promoting mental wellbeing in public health policy (Faculty of Public Health/Mental Health Foundation, 2016; Walker et al., 2019; Welsh Government, 2012). Wellbeing has also become the focus of a national and local government policy agenda focused on sustainable development and alternative measures of societal and economic progress beyond health outcomes (All-Party Parliamentary Group on Wellbeing Economics, 2014; Wellbeing of Future Generations (Wales) Act, 2015; Welsh Government, 2021b).

Despite a long history of the WHO defining health and mental health as states of wellbeing (WHO, 2004, 2020), the terms health and mental health are frequently used to denote illness and health (and mental health) services (Dooris et al., 2017; Stewart-Brown, 2018; Tudor, 1996). Wellbeing is now often the term used in practice to denote positive health and mental health (Dooris et al., 2017; Faculty of Public Health/Mental Health Foundation, 2016; Stewart-Brown, 2018). Atkinson (2011) articulates neatly some of the dilemmas for the public health field in defining and measuring wellbeing: aligning subjective wellbeing measures such as life satisfaction to mental health may create opportunities for greater investment in interventions to improve population mental health. On the other hand, subjective wellbeing measures represent wellbeing at an individual – and subjective - level only, leading to a focus on individually targeted responses. This may detract from action on the social determinants of health and reducing inequalities, and hamper efforts to engage cross-sector partners in steering policy towards improving wellbeing.

Keeping a broad approach to wellbeing as a "*unifying concept across different sectors*" to be addressed beyond the confines of the health and medical domains is ultimately seen by Atkinson (2011) as more useful. A recent paper by the WHO (2021) appears to agree, identifying that "*wellbeing concepts can bring policy coherence across sectors and galvanize action*". In an example of such policy coherence, Wales has put in place legislation for a cross-governmental approach to wellbeing and sustainable development in the Wellbeing of Future Generations (Wales) Act (2015) and a set of national indicators that monitor progress in improving wellbeing at a population level. These include measures of wellbeing at individual (mean mental wellbeing), community (people feeling that they belong to their local area), cultural (Welsh language speakers), economic (people in employment), and planetary levels (healthy ecosystems) (Welsh Government, 2021b).

Health (and Wellbeing) in All Policies

Governments and other public and private organisations make decisions and investments, design programmes and services, and develop policies that are likely to impact on the health and wellbeing of the population. As the WHO Commission on Social Determinants of Health (CSDH) identified: "*Health and health equity may not be the aim of all social policies but they will be a fundamental result*" (CSDH, 2008, p. 1). The definitions and frameworks of health and wellbeing utilised to interpret impacts and inform decisions are therefore important because they will ultimately shape the design of policies/programmes and what type of responses (if any) are developed. If a deficit or illness model is used, the responses are likely to be defined in terms of health service provision. If health is interpreted as physical health, then the policy responses may not consider impacts on mental wellbeing. If wellbeing is only viewed at an individual level, opportunities for investment in building community assets and addressing structural or systemic factors may not be within scope.

To enable health and wellbeing to be central considerations in policy- and decision-making, the health promotion field developed "Health in All

Policies" (HiAP), which is a systematic approach to maximising opportunities and minimising harms for health and wellbeing arising from cross-sector policies and decisions (WHO, 2014). Health impact assessment (HIA) is recognised as a key tool for achieving HiAP (Green et al., 2021) and achieves this by the application of multilevel frameworks of health, wellbeing and equity to policy proposals in a process of active engagement with policy- and decision-makers. Participatory HIAs can also give citizens, communities, and/or service users a stronger voice in decision-making (Dannenberg, 2016); therefore, the process of HIA can also benefit health and wellbeing via enhancement of autonomy. Harris-Roxas and Harris (2011) provide a typology of HIAs, and the examples in Box 8.3 below demonstrate the varied and rich opportunities for the application of HIA.

Box 8.3 Typology of HIAs (Harris-Roxas & Harris, 2011) with Examples from Practice

- *Mandated HIA*: undertaken to fulfil a statutory or regulatory requirement, usually linked to environmental legislation and for major economic, transport and infrastructure development. Example: HIA on Wylfa Newydd Nuclear Power Station in North Wales (Horizon Nuclear Power, 2018).
- *Decision-support HIA*: conducted voluntarily by, or with the agreement of, organisations responsible for a policy or programme. Example: HIA on Supplementary Planning Guidance for open space in a local authority, which determines the type and amount of open space provided in a community (Green & Duncalf, 2017).
- *Advocacy HIA*: aim to influence policy and decision-making by ensuring that under-recognised health and wellbeing concerns are addressed. Example: a HIA on the proposed Trans Pacific Partnership Agreement on international free trade (Hirono et al., 2016).
- *Community-led HIA*: empower communities to influence decisions that may affect their health by ensuring that the community's health-related concerns are recognised and addressed. Example: a community-led HIA on fossil fuel extraction in South Wales (Golby & Lester, 2005).

Evaluation and reviews of HIAs find that the outcomes of the application of HIA include influencing policy decisions, improved collaboration and partnership working, increased awareness of health and wellbeing issues among decision-makers, and increased community participation in decision-making (Dannenberg, 2016; Sohn et al., 2018). Importantly, Sohn et al. (2018) also found that HIA can influence "*the systems and processes that determine the distribution of health promoting resources*", highlighting the role that HIAs can play in systemic action on health and wellbeing.

Mental Wellbeing Impact Assessment (MWIA)

Mental Wellbeing Impact Assessment (MWIA) is a HIA tool that identifies impacts specifically on mental wellbeing of policies, programmes, services, and projects (Cooke & Stansfield, 2009). It has a "salutogenic" approach by assessing impacts on factors that promote and protect mental wellbeing and has an emphasis on addressing health inequalities by focusing on those most at risk of poor mental wellbeing (Cooke et al., 2011). MWIA shares the same process, values and principles, and methods as HIA (Cooke & Stansfield, 2009). The MWIA Toolkit (Cooke et al., 2011) does not define wellbeing or mental wellbeing; instead, the process of MWIA encourages participants to create a definition of mental wellbeing that is meaningful to them and grounded in the context of the people and proposal that is under assessment. It is framed around three core "protective factors" and the social determinants of mental health and wellbeing, with subcomponents at individual, community, and structural levels (see Box 8.4 for a summary). The MWIA framework conceptualises wellbeing as operating at individual, community, and societal levels and reflects hedonic ("emotional wellbeing") as well as eudaimonic perspectives (Westerhof & Keyes, 2010) including components of "self-determination", "feeling valued", and "learning and development". Physical wellbeing is also addressed via components of "healthy lifestyles" and "physical environment" such as access to green and natural spaces.

Box 8.4 Protective Factors for Mental Wellbeing Used in MWIA (Cooke et al., 2011)

- *Enhancing control*: e.g. a sense of control, self-determination, influencing decisions.
- *Increasing resilience and community assets*: e.g. emotional wellbeing, cognitive and social skills, beliefs and values, social relationships and networks, shared public spaces, sustainable local economy.
- *Facilitating participation and inclusion*: e.g. a sense of belonging, having a valued role, community cohesion.
- *Social determinants of mental wellbeing*: e.g. economic security, housing security and quality, physical environment, education, good quality food, and tackling discrimination.

MWIA is used internationally and had been applied in a wide range of contexts, policy sectors and programmes (see Box 8.5) representing its broad acceptability and applicability.

Box 8.5 Examples of MWIA in Different Contexts and Population Groups

- Arts programmes. Example: MWIA on arts programme for older adults (Shearn, 2011).
- Workplace wellbeing. Example: MWIA used to inform actions to improve mental wellbeing in the NHS and local government (Burford et al., 2017).
- Natural disasters. Example: MWIA on community recovery from a tsunami on Robinson Crusoe Island, Chile (Ampuero et al., 2015).
- Nature-based interventions. Example: MWIA used to evaluate a therapeutic horticulture programme with children (Chiumento et al., 2018).
- Adult education. Example: MWIA on a women's community education course (Lewis et al., 2013).

Application of MWIA tool shares similar impacts and outcomes with HIA, although evaluation of MWIA is less well developed. Evaluation to date includes case study evidence (Burford et al., 2017; Cooke & Stansfield, 2009) and internal evaluation of larger scale MWIA programmes (e.g. Edmonds, 2009; West et al., 2007). A first-level review of the impact and outcomes concluded that MWIA achieves "*its objectives of orientating initiatives towards supporting the evidenced protective factors for wellbeing*" (King, 2014 p. 8). Few MWIAs appear in peer-reviewed academic journals, and this is a limitation in the current field of practice, research, and development.

The review of case studies of MWIA (Cooke & Stansfield, 2009) found that the application can support the development of shared understandings and coherent approaches to mental wellbeing with a range of stakeholders and enable policies, programmes, and projects to have a positive impact on mental wellbeing. Importantly, it provides a process that:

> *enables a shift in thinking and resources to improving wellbeing. It will contribute to shifting systems from those that concentrate on managing the consequences of poor wellbeing (high crime, unemployment, illness, intolerance, under achievement) to ones that tackle its determinants: control, resilience, participation and inclusion* (p. 4).

Therefore, in common with HIA, the process and framework used in MWIA have demonstrated a capacity to influence system-level thinking and actions on improving mental wellbeing.

Expanding the Role of the HIA and MWIA in Addressing Complex Societal Challenges

In 2018/19 the Wales HIA Support Unit conducted a HIA on the implications of "Brexit" for public health in Wales (Green et al., 2019). The HIA on "Brexit" was "*unique*" in that it did not focus on a single policy or programme, which is usual practice in HIA, but the totality of implications of the UK leaving the European Union (EU) (Green, Ashton, et al., 2020). The HIA identified pathways through which there could be direct and indirect impacts on health and wellbeing. For example, indirect impacts on mental wellbeing were identified as possible via changes to economic conditions, uncertainty, and feeling included (for example, for

EU citizens living in the UK) (Green et al., 2019). Possible direct impacts on environmental and working conditions via legal and regulatory change were also identified, along with reductions in EU funding to address social, economic, and health inequalities in Wales (Green et al., 2019). The analysis and findings in the HIA enabled priorities for population health and wellbeing in relation to "Brexit" to be identified in Wales, and the majority of recommendations were implemented to support planning (Green, Ashton, et al., 2020). For example, the HIA led to the development of work programme to understand the implications of future trade policy on health and wellbeing (Green, Ashton, et al., 2020); investment in research on the health and wellbeing impacts of "Brexit" on population groups identified as potentially disproportionally impacted by "Brexit" in Wales such as farmers (Davies et al., 2020); fishers (Homolova et al., 2020); and young people (Petchey et al., 2020), and was used to inform national strategic planning (Green, Ashton, et al., 2020).

The HIA on Brexit led to recognition of the utility of HIA as a method for analysing novel, complex, strategic, and uncertain scenarios that have major implications for population health and wellbeing and has since resulted in the commissioning of a HIA on climate change in Wales (Green, Ashton, et al., 2020); a series of HIAs on the wider health, wellbeing, and equity impacts of the COVID-19 pandemic (Green, Ashton, et al., 2020; Green, Morgan, et al., 2020); and an MWIA on the impact of the COVID-19 pandemic on the mental wellbeing of young people (Edmonds, Morgan et al. 2022). Stakeholder engagement has been an important component of each of these impact assessments, enabling a broad range of insights on how these developing societal challenges may influence health and wellbeing in Wales and the potential areas for multisectoral action.

Conclusion

This chapter has placed wellbeing in the context of improving population health and wellbeing, and explored some areas of progress and ongoing challenges. The public health tools described in this chapter (HIA and MWIA) have demonstrated outcomes including the development of shared understandings and coherent approaches to wellbeing, shifting systems to focus on the determinants of wellbeing, and influencing the distribution of health and wellbeing promoting resources. Synergies between the proposed "third wave" of positive psychology (Lomas et al., 2020; Mead et al., 2021) and the "fifth wave" of public health (Hanlon et al., 2011) appear to offer common ground ripe for further exploration and mutual learning.

References

All-Party Parliamentary Group on Wellbeing Economics (2014). *Wellbeing in four policy areas.* Retrieved March 21, 2022, from appg-report-2014.pdf (wellbeingeconomics.co.uk).

Ampuero, D., Goldsworthy, S., Delgado, L., & Miranda, J. C. (2015). Using mental well-being impact assessment to understand factors influencing well-being after a disaster. *Impact and Project Appraisal, 33*(3), 184–194. https://doi.org/10.1080/14615517.2015.1023564

Atkinson, S. (2011). Moves to measure wellbeing must support a social model of health. *British Medical Journal, 343*, d7323. https://doi.org/10.1136/bmj.d7323

Bartlett, C. J., & Coles, E. C. (1998a). Psychological health and well-being: Why and how should public health specialists measure it? Part 1: Rationale and methods of the investigation, and review of psychiatric epidemiology. *Journal of Public Health Medicine, 20*(3), 281–287.

Bartlett, C. J., & Coles, E. C. (1998b). Psychological health and well-being: Why and how should public health specialists measure it? Part 2: Stress, subjective well-being and overall conclusions. *Journal of Public Health Medicine, 20*(3), 288–294.

Berghs, M., Atkin, K., Hatton, C., & Thomas, C. (2019). Rights to social determinants of flourishing? A paradigm for disability and public health research and policy. *BMC Public Health, 19*, 997. https://doi.org/10.1186/s12889-019-7334-8

Buck, D., Baylis, A., Dougall, D., & Robertson, R. (2018). *A vision for population health: Towards a healthier future.* The Kings Fund. Retrieved March 15, 2022, from A vision for population health | The King's Fund (kingsfund.org.uk).

Burford, C., Davey, S., Knight, A., King, S., Cooke, A., & Coggins, T. (2017). Mental well-being impact assessment in the workplace. *Journal of Public Mental Health, 16*(3), 99104–99112. https://doi.org/10.1108/JPMH-01-2017-0002

de Cates, A., Stranges, S., Blake, A., & Weich, S. (2015). Mental well-being: An important outcome for mental health services? *The British Journal of Psychiatry, 207*, 195–197. https://doi.org/10.1192/bjp.bp.114.158329

Chiumento, A., Mukherjee, I., Chandna, J., Dutton, C., Rahman, A., & Bristow, K. (2018). A haven of green space: Learning from a pilot pre-post evaluation of a school-based social and therapeutic horticulture intervention with children. *BMC Public Health, 18*, 836. https://doi.org/10.1186/s12889-018-5661-9

Cooke, A., & Stansfield, J. (2009). *Improving mental well-being through impact assessment: A summary of the development and application of a mental well-being impact assessment tool.* National Mental Health Development Unit/Department of Health (England).

Cooke, A., Friedli, L., Coggins, T., Edmonds, N., Michaelson, J., O'Hara, K., Snowden, L., Stansfield, J., Steuer, N., & Scott-Samuel, A. (2011). *Mental well-being impact assessment: A toolkit for well-being* (3rd ed). Members of the National MWIA Collaborative (England). Retrieved March 15, 2022, from (PDF) Mental Well-being Impact Assessment A Toolkit for Well-being, Members of the National MWIA Collaborative (England) (researchgate.net).

CSDH. (2008). *Closing the gap in a generation: Health equity through action on the social determinants of health. Final report of the commission on social determinants of health.* World Health Organization. Retrieved March 20, 2022, from Closing the gap in a generation: health equity through action on the social determinants of health (who.int).

Dannenberg, A. L. (2016). Effectiveness of health impact assessments: A synthesis of data from five impact evaluation reports. *Preventing Chronic Disease, 13*, 150559. https://doi.org/10.5888/pcd13.150559

Davies, A. R., Grey, C. N. B., Homolova, L., Fisher, J., Burchett, N., & Kousoulis, A. (2020). *An action framework to support the mental health and well-being of farmers at times of uncertainty.* Retrieved March 31, 2022, from Farmers' health and wellbeing needs to be at the centre of Brexit response – new report - Public Health Wales (nhs.wales).

Dooris, M. T., Farrier, A., & Froggett, L. (2017). Wellbeing: The challenge of 'operationalising' an holistic concept within a reductionist public health programme. *Perspectives in Public Health, 138*(2), 93–99. https://doi.org/10.1177/1757913917711120

Edmonds, N. (2009). *Well London: Mental wellbeing impact assessment project report.* Retrieved March 20, 2022, from (PDF) Well London: Mental Well Being Impact Assessment Project Report (researchgate.net).

Edmonds, N., Morgan, L., Thomas, H. A., Fletcher, M., Parry Williams, L., Evans, L., Green, L., Azam, S., Bellis, M. A. (2022). Protecting the mental wellbeing of our future generations: learning from COVID-19 for the long term. A Mental Wellbeing Impact Assessment Approach. Main Findings Report. Public Health Wales NHS Trust. Retrieved July 30th 2022 Protecting the mental wellbeing of our future generations: learning from COVID-19 for the long term - World Health Organization Collaborating Centre On Investment for Health and Well-being (phwwhocc.co.uk)

Edwards, D. J., Rainey, E., Boukouvala, V., et al. (2019). Novel ACT-based eHealth psychoeducational intervention for students with mental distress: a study protocol for a mixed-methodology pilot trial BMJ Open 2019;9:e029411. https://doi.org/10.1136/bmjopen-2019-029411

Faculty of Public Health/Mental Health Foundation (2016). *Better mental health for all: A public health approach to mental health improvement*. Retrieved March 20, 2022, from better-mental-health-for-all-final-low-res.pdf (fph.org.uk).

Golby, A., & Lester, C. (2005). *Health impact assessment of the proposed extension to Margam opencast mine*. Retrieved March 20, 2022, from Kenfig draft latest Dec 05.doc (phwwhocc.co.uk).

Green, L., & Duncalf, L. (2017). *Rapid health impact assessment (HIA) of draft recreational public open space supplementary planning guidance (SPG)*. Retrieved March 20, 2022, from Microsoft Word - DCC Open Space SPG HIA final report - Feb 2017.docx (phwwhocc.co.uk).

Green, L., Edmonds, N., Morgan, L., Andrew, R., Ward, M., Azam, S., & Bellis, M. A. (2019). *The public health impact of Brexit in Wales: A health impact assessment approach. Main findings*. Retrieved March 22, 2022, from The Public Health Implications of Brexit in Wales: A Health Impact Assessment Approach - World Health Organization Collaborating Centre On Investment for Health and Well-being (phwwhocc.co.uk).

Green, L., Ashton, K., Edmonds, N., & Azam, S. (2020). Process, Practice and Progress: A case study of health impact assessment of Brexit in Wales. *International Journal of Environmental Research and Public Health, 17*(18), 6652. https://doi.org/10.3390/ijerph17186652

Green, L., Morgan, L., Azam, S., Evans, L., Parry-Williams, L., Petchey, L., & Bellis, M. A. (2020). *A health impact assessment of the "Staying at Home and Social Distancing Policy" in Wales in response to the COVID-19 pandemic*. Retrieved March 22, 2022, from A Health Impact Assessment of the 'Staying at Home and Social Distancing Policy' in Wales in response to the COVID-19 pandemic. - World Health Organization Collaborating Centre On Investment for Health and Well-being (phwwhocc.co.uk).

Green, L., Ashton, K., Bellis, M. A., Clemens, T., & Douglas, M. (2021). 'Health in all policies'—A key driver for health and well-being in a post-COVID-19 pandemic world. *International Journal of Environmental Research and Public Health, 18*, 9468. https://doi.org/10.3390/ijerph18189468

Hanlon, P., Carlisle, S., Hannah, M., Reilly, D., & Lyon, A. (2011). Making the case for a 'fifth wave' in public health. *Public Health, 125*(1), 30–36. https://doi.org/10.1016/j.puhe.2010.09.004

Harris-Roxas, B., & Harris, E. (2011). Differing forms, differing purposes: A typology of health impact assessment. *Environmental Impact Assessment Review, 31*(4), 396–403. https://doi.org/10.1016/j.eiar.2010.03.003

Hayes, S. C., Strosahl, K. D., & Wilson, K. G. (2011). *Acceptance and Commitment Therapy, Second Edition: The Process and Practice of Mindful Change*. New York, NY: The Guilford Press.

Hirono, K., Haigh, F., Gleeson, D., Harris, P., Thow, A. M., & Friel, S. (2016). Is health impact assessment useful in the context of trade negotiations? A case study of the Trans Pacific Partnership Agreement. *BMJ Open, 6*(4), e010339. https://doi.org/10.1136/bmjopen-2015-010339

Homolova, L., Grey, C. N. B., Burchett, N., & Davies, A. R. (2020). *Building resilience in the fishing sector.* Retrieved March 31, 2022, from https://phw.nhs.wales/publications/publications1/building-resilience-in-the-fishing-sector-in-wales-english/.

Horizon Nuclear Power (2018). *Wylfa Newydd Project 8.19: Health impact assessment report.* Retrieved March 21, 2022, from EN010007-001739-8.19 Health Impact Assessment Report (Rev 1.0).pdf (planninginspectorate.gov.uk).

Johnson, R., Jenkinson, D., Stinton, C., Taylor-Phillips, S., Madan, J., Stewart-Brown, S., & Clarke, A. (2016). Where's WALY?: A proof of concept study of the 'wellbeing adjusted life year' using secondary analysis of cross-sectional survey data. *Health and Quality of Life Outcomes, 14*(1), 126. https://doi.org/10.1186/s12955-016-0532-5

Kemp, A. H., & Fisher, Z. (2022). Wellbeing, whole health and societal transformation: Theoretical insights and practical applications. *Global Advances in Health and Medicine* 11. https://doi.org/10.1177/21649561211073077

Kern, M. L., Williams, P., Spong, C., Colla, R., Sharma, K., Downie, A., Taylor, J. A., Sharp, S., Siokou, C., & Oades, L. G. (2019). Systems informed positive psychology. *The Journal of Positive Psychology,* 1–11. https://doi.org/10.1080/17439760.2019.1639799.

King, S. (2014). *How effective is mental well-being impact assessment? An evaluation by the Tavistock Institute of Human Relations.* Retrieved March 22, 2022, from How Effective is Mental Well-being Impact Assessment?—The Tavistock Institute (tavinstitute.org).

Layard, L., & Oparina, E. (2021) Living long and living well*: The WELLBY approach.* In J. F. Helliwell, R. Layard, J. D. Sacks, J. De Neve, L. B. Aknin, & S. Wang (Eds.), *World Happiness Report.* Retrieved May 9, 2022, from World-Happiness-Report-2021.pdf (researchgate.net).

Lewis, L., Henswon, J., Wood, A., & Chapman, L. M. (2013). *WEA community health education in Stoke-on-Trent project: Women's Boot Camp, Cobridge MWIA.* Retrieved March 15, 2022, from WEA Community Health Education in Stoke-on-Trent (CHEST) project: Women's Boot Camp, Cobridge on Mental Well-being Impact Assessment (MWIA) (healthhumanities.org).

Lomas, T., Waters, L., Williams, P., Oades, L. G., & Kern, M. L. (2020). Third wave positive psychology: Broadening towards complexity. *The Journal of Positive Psychology, 16,* 660.

Marmot, M., Allen, J., Goldblatt, P., Boyce, T., McNeish, D., Grady, M., & Geddesm I. (2010). *Fair society, healthy lives—The Marmot review.* Institute of Health Equity. Retrieved May 9, 2022, from Fair Society Healthy Lives full report (parliament.uk).

Mead, J., Fisher, Z., & Kemp, A. H. (2021). Moving beyond disciplinary silos towards a transdisciplinary model of wellbeing: An invited review. *Frontiers in Psychology, 12*, 642093. https://doi.org/10.3389/fpsyg.2021.642093

Mehta, N., Croudace, T., & Davies, S. C. (2015). Public mental health: Evidenced-based priorities. *The Lancet, 385*, 1472–1475. https://doi.org/10.1016/s0140-6736(14)61400-8

National Institute for Health and Care Excellence (2012). *The guidelines manual: Section 7: Assessing cost effectiveness.* Retrieved March 20, 2022, from 7 Assessing cost effectiveness | The guidelines manual | Guidance | NICE.

Petchey, L., Davies, A., Urbano, S., Azam, S., & Davies, A. (2020). *The health and well-being of young people in Wales: Do they think Brexit will make a difference?* Retrieved March 31, 2022, from Brexit_young_people_report_English_final_Feb_2020-1.pdf (phwwhocc.co.uk).

Public Health Wales Observatory (2020). *Mental wellbeing in Wales.* Retrieved March 20, 2022, from Mental Wellbeing in Wales (2020) - Public Health Wales (nhs.wales).

Senior, S. L., Caan, W., & Gamsu, M. (2020). Welfare and well-being: Towards mental health-promoting welfare systems. *The British Journal of Psychiatry, 216*(1), 4–5. https://doi.org/10.1192/bjp.2019.242

Shearn, H. (2011). *Art into life for older adults with mental health problems and dementia* MWIA. Retrieved March 22, 2022, from MWIA ART INTO LIFE FINAL REPORT MAY 2011 (artshealthresources.org.uk).

Sohn, E. K., Stein, L. J., Wolpoff, A., Lindberg, R., Baum, A., Mc-Innis Simoncelli, A., & Pollack, M. (2018). Avenues of influence: The relationship between health impact assessment and determinants of health and health equity. *Journal of Urban Health, 95*, 754. https://doi.org/10.1007/s11524-018-0263-5

Stewart-Brown, S. (2018). Resilience and well-being. In D. Bhugra, K. Bhui, S. Y. S. Wong, & S. E. Gilman (Eds.), *Oxford Textbook of Public Mental Health.* Oxford University Press.

Stewart-Brown, S., Middleton, J., & Ashton, J. (2015). Responses to the Chief Medical Officer's report 2013. *The Lancet, 385*(9987), 2576. https://doi.org/10.1016/s0140-6736(15)61157-6

Tudor, K. (1996). *Mental health promotion: Paradigms and practice.* Routledge.

Wahlbeck, K. (2015). Public mental health: The time is ripe for translation of evidence into practice. *World Psychiatry, 14*(1), 36–42. https://doi.org/10.1002/wps.20178

Walker, I. F., Stansfield, J., Makurah, L., Garnham, H., Robson, C., Lugton, C., Hey, N., & Henderson, G. (2019). Delivering national public mental health—Experience from England. *Journal of Public Mental Health, 18*(2), 112–123. https://doi.org/10.1108/JPMH-06-2018-0032

Wanless, D. (2003). *Securing good health for the whole population: Final report.* HM Treasury. Retrieved March 15, 2022, from HMTreasury amends.pdf (vcu.edu).

Well-being of Future Generations (Wales) Act (2015). Retrieved March 15, 2022, from Well-being of Future Generations (Wales) Act 2015 (legislation.gov.uk).

Welsh Government (2012). *Together for mental health: A strategy for mental health and wellbeing in Wales*. Retrieved March 22, 2022, from DRAFT v3 (gov.wales).

Welsh Government (2021a). *Wellbeing of Wales*. Retrieved March 20, 2022, from Wellbeing of Wales: 2021 | GOV.WALES.

Welsh Government (2021b). *Well-being of future generations act: The essentials*. Retrieved March 21, 2022, from Well-being of future generations act: the essentials | GOV.WALES.

West, H., Hanna, J., Scott-Samuel, A., Cooke, A. (2007). *Liverpool 08 European capital of culture: Mental well-being impact assessment*. Retrieved March 21, 2022, from (PDF) Liverpool 08 European Capital of Culture: Mental Wellbeing Impact Assessment (researchgate.net).

Westerhof, G. J., & Keyes, C. L. M. (2010). Mental illness and mental health: The two continua model across the lifespan. *Journal of Adult Development, 17*, 11–119. https://doi.org/10.1007/s10804-009-9082-y

World Health Organization (2004). *Promoting mental health: Concepts, emerging evidence, practice*. Retrieved March 20, 2022, from http://www.who.intlmen-talhealth/evidence/en/promoting mhh.pdf.

World Health Organization (2009). *Milestones in health promotion: Statements from Global Conferences*. Retrieved March 15, 2022, from Milestones in Health Promotion: Statements from Global Conferences (who.int).

World Health Organization (2014). *Health in all policies: Helsinki statement. Framework for country action*. Retrieved March 20, 2022, from Health in all policies: Helsinki statement. Framework for country action (who.int).

World Health Organization (2020). *Basic documents 49th edition*. Retrieved March 21, 2022, from BASIC DOCUMENTS (who.int).

World Health Organization (2021). *Towards developing WHO's agenda on well-being*. Retrieved March 20, 2022, from Towards developing WHO's agenda on well-being.

World Health Organization (2022). *10 essential public health operations*. Retrieved March 20, 2022, from WHO/Europe | Public health services - The 10 Essential Public Health Operations.

CHAPTER 9

The Economics of Individual Wellbeing and the Transformation of Society

Don J. Webber and Dominic Page

Abstract The mainstream economics stance focusing on how to allocate government resources to enhance wellbeing and alleviate poor mental health is fundamentally and inherently misguided because it is underpinned by the discipline's staunch assumptions that the most effective best way to model the economy is to assume that people are autonomous self-interested individuals. This mathematically motivated, peer-driven approach generates clear, simple, and highly respected numerical results which are needed in a world that prioritises evidence-based decision making. Unfortunately for the modellers, our world is not full of autonomous self-interested individuals but socialised other-regarding individuals. The

D. J. Webber (✉)
Sheffield University Management School, The University of Sheffield, Sheffield, UK
e-mail: d.j.webber@sheffield.ac.uk

D. Page
Gloucestershire Business School, University of Gloucestershire, Cheltenham, UK
e-mail: dpage@glos.ac.uk

A. H. Kemp, D. J. Edwards (eds.), *Broadening the Scope of Wellbeing Science*, https://doi.org/10.1007/978-3-031-18329-4_9

powerful economic discipline's key modelling assumptions reflect a poor understanding of the networked socio-spatial fabric of society that is central to the social model of ill-health. After delving into the structural limitations of mainstream economics, we call for a seismic shift in thinking in the economics profession so that it more effectively embraces social networks, sociological, and spatial factors. Doing so would enable a more appropriate allocation of resources that could more effectively transform the social system.

Keywords Institutional economics • Resource allocation • Societal development • Policy recommendations

Introduction

Many societies possess an aim to improve the wellbeing of their inhabitants. Individuals with higher levels of wellbeing are not only more productive but also experience lower levels of physical and mental ill-health. How can society achieve improvements in wellbeing and where and when should resources be allocated to achieve this aim? Economists have a focus on resource allocation decisions and are venerated by policymakers for grounding their policy recommendations on hard numerical evidence. What is much less appreciated is that the dominant economics narrative leads to a constrained set of results and hence policy recommendations, as the narrative is built on discipline-engrained implicit ontological, epistemological, and axiological foundations, assembled within profit and utility maximisation assumptions, and dependent on data that fit this narrative.

Most mainstream economists base their arguments on the assumption that firms focus their attention on profit (or revenue) maximisation. This habitually adopted restrictive assumption is made for the sake of mathematical modelling convenience, and why would a firm consciously choose to avoid maximising its profits? The argument then proceeds that if the benefits for firms from applying policies, practices, and interventions that enhance the wellbeing of their workers are less than the costs, then the incentive won't be adopted. A similar argument is made when dealing with individuals: any individual is assumed to have the freedom to make choices that maximise their life satisfaction and wellbeing, and why would anyone consciously choose to avoid doing something that increases their satisfaction or enhances their level of wellbeing? In a laissez-faire setting where employees maximise their satisfaction and wellbeing and where firms maximise their profits, we can tautologically deduce that an employee's wellbeing is positively associated with their ability to contribute to the productivity of their organisation. Hence, leaving individuals to their own devices will not only maximise their satisfaction and wellbeing but will also maximise productivity and wellbeing and enable society to move up the development ladder.

Although individuals have a fundamental personal responsibility for maintaining their health (Black, 2008, p. 15), it is important to recognise that those same individuals are institutionally clustered within families, workplaces, communities, conurbations, and nations within and across our socialised planet. This chapter emphasises that this institutionalised social clustering (which is socially, temporally, and spatially dependent within an evolving path dependent context) is fundamental to people's wellbeing but is excluded from the abstract empirical modelling approach of most mainstream economists; this omission critically undermines economists' models, making them spuriously accurate at best and not even wrong[1] at worst. The focus of much economic policy on enhancing profitability, efficiency, and/or productivity creates a problem if society is concerned with the wellbeing of its inhabitants: as our economy is rated on outputs, typically measured using gross value added, greater quantities of drugs that are produced and sold to cure individuals' wellbeing problems would indicate that society is going in the right direction, but the measurement of outputs might reflect the level of need to cure wellbeing problems created by the system rather than an ability to prevent low levels of wellbeing.

This chapter argues that modelling an individual's wellbeing with under-identified interpersonal connections will create inaccurate and biased results. Although the dominant areas of the economics profession are commendably focused on using resources efficiently, fervent on the application of quantitative methods, and powerful at producing an evidence-base on which to inform government policy, it is vital to recognise the discipline's inherent limitations, not only because of the inadequate contemporary appreciation of the depth and importance of social issues but also because much of the mainstream economics literature

[1] Myddelton (2009) defines spurious accuracy as a pretence to precision that is either unattainable or useless (or both). Woit (2006) argues that a scientific idea is "not even wrong" if it is so incomplete that it cannot be used to make predictions that could be compared to observations to see if the idea is wrong.

remains steadfast to a closed set of ontological, epistemological, and axiological positions that underpin its analytical approach and policy recommendations.

Other-Regarding or Self-Interest?

The prevailing group of economists (sometimes called mainstream neoclassical economists) typically estimate models that implicitly assume individuals behave "as if" they are autonomous self-interested rational decision-makers and make choices that generate maximum levels of wellbeing and satisfaction. This constrained optimisation approach is a natural result of the ontological, epistemological, and axiological foundations subconsciously adopted by this influential group. This predisposition builds on formal university methods training in economics, which is typically constrained only to econometrics. Webber and Pattinson (2022) emphasise that peer group forces relegate the primacy of other methods, stimulate a continuous need to update econometric knowledge, and make it necessary to showcase these econometric abilities to their peers, meaning that it is often the conspicuous showcasing of specific methodological abilities that matters rather than the substantive contribution to knowledge.[2]

Estimation of econometric models requires the provision of large datasets, and the large number of individuals in many official statistics datasets adds a veil of rigour and respectability, often making the findings difficult to question by the already converted or those untrained in the limits of econometrics. These datasets tend to draw on large-scale surveys across thousands of residents and claim to be nationally representative. For instance, the UK's Office for National Statistics provides access to datasets that contain answers to wellbeing related questions on a Likert scale of 0 (zero) to 10, where 0 is "not at all" and 10 is "completely". The *life satisfaction* measure is the response to the question: "Overall, how satisfied are you with your life nowadays?", the *worthwhile* measure is the response to the question, "Overall, to what extent do you feel that the things you do

[2] Econometric papers score highly in research assessment exercises where, of course, the journal market and hence rankings, are socially determined. "The market is not a natural phenomenon… The observation that the market is a manmade institution rather than a natural phenomenon is simple enough. But most of what passes as economic wisdom is based on a denial of that historical truth" (Dugger, 1989, p. 607).

in your life are worthwhile?", the *happiness* measure is the response to the question: "Overall, how happy did you feel yesterday?", and the *anxiety* measure is the response to the question: "Overall, how anxious did you feel yesterday?". Analyses of these datasets implicitly assume that the individuals contained in the dataset optimise life satisfaction, worthiness, and happiness while minimising anxiety. Critically, health is conceptualised as an "individual" experience. Although these estimations undoubtedly have worth, it must be recognised that social networks, relative positioning, status, and relative and absolute poverty are not included in their models. These models do not permit health and wellbeing to be affected by social forces that evolve. Social networks and the development, structure, and functioning of human society, which cluster in space, are either assumed not to be important in the estimation of the wellbeing model or are assumed to be part of a well-behaved error term. Thus, implementing the dominant economic approach completely ignores the rise in awareness of the social determinants of poor mental health and wellbeing, and instead conceives wellbeing as an individual pathology or medical problem to be treated. The associated assumption that mental illnesses are "ultimately defined by a sub-optimal social functioning" (Pilgrim & Rogers, 2003, p. 4) has been increasingly challenged, not least as an explanation for economic disadvantage. As highlighted by Barnes et al. (2000), the inclusion of mental health within a framework of "medical conditions" is problematic on conceptual grounds: it obscures the reality of the social impact on mental health and the experiences of power and powerlessness.

There Is an Alternative

Institutional economists reject mainstream economic thought primarily because they do not believe that every economic phenomenon should be analysed from the abstract ("as-if") concepts of maximisation and equilibrium. Instead, institutional economists conceive that a substantial and dominant part of human behaviour and wellbeing is influenced by what others do, what an individual perceives others do, or how one person relates to or is positioned relative to others. Hence, the institutional economic perspective assumes people are embedded in a community of people who consciously (or subconsciously) affect each other's preferences, behaviours, and wellbeing.

Adam Smith, known as the father of economics, wrote his first book, *The Theory of Moral Sentiments* (1759). In it, he emphasised that people seek the approval of their impartial spectator, who is the person in the mirror—their internal conscience—and whose disapproval makes us aware of the nature of our conduct. Adam Smith emphasised the socially informed need for us to feel connections with the passions of others; he had a key focus on how sympathy informs these moral sentiments. Part of Adam Smith's work was in response to Hume's (1739-40) argument that principles of human nature are universal and unchanging and from which social institutions and social behaviours can be deduced, i.e. a person deduces from others the roles of institutions and expected interpersonal behaviours in shaping their own behaviours. But Adam Smith argues that "[T]he jurisdiction of the man within is founded altogether on the desire of praise-worthiness and in the aversion to blame-worthiness" which underlies our "dread of possessing those qualities, and performing those actions, which we hate and despise in other people" (Smith, 1959, iii.2.33). Indeed, Vernon Smith (Economics Nobel Prize winner from 2002) recognises that Adam Smith's two most famous books encompass not only the trading of goods but also gifts, assistance, and favours out of sympathy that are self-interested *and* other-regarding. However, much of the modern-day mainstream economics discipline appears to have unlearnt or disregarded Adam Smith's contributions on the grounds of moral sentiment.

Prior to World War II, there was a wide range of economic thought, but economists began to unlearn or move away from the analysis of the importance of social connections after WWII. Harvey (2016) explains why this happened: WWII created conditions of full employment in the US where applications of quantitative marginal economic analyses excel. During the Cold War, economics professors in Western economies that were perceived to be sympathetic towards Russian and Chinese economic policies faced sanctions from their universities, funding agencies, governments, and private sector interests, and were ostracised. This reduced the breadth of economic thinking and elevated the importance of estimating models using Marshallian marginal analyses, thereby favouring the recruitment and promotion of neoclassical economists. To this day, socially informed journal rankings (and hence promotion) remain in this line of thought and few neoclassical economists are aware of the existence or strengths of other communities of economics scholars (Harvey, 2016).

Social Ties in the Evidence Base

Consistent with Adam Smith's *other-regarding* moral sentiments, Mind's Five Ways to Wellbeing stresses the importance of interpersonal connections and giving within a community. However, modern mainstream economics focuses on the level of the *self-interested* individual. Mainstream economists typically analyse official statistics that ask questions consistent with the self-interested individual, perhaps guided by economic thought. Unfortunately, these official datasets very rarely include information on social ties, social positioning, or social status even though they are crucial determinants of behaviour. This creates an "undersocialized account" that atomises human behaviour (Granovetter, 1985) and separates the mainstream economic analysis from institutional economics.

Veblen's (1899) focus on consumption to engender an image of social status has a strong following in institutional economics. Veblen argued that as others conspicuously consume to illustrate their social status, so an individual is compelled to also consume conspicuously to illustrate that their status is at least on a par with if not better than those around them; this process is nicknamed "Keeping up with the Joneses." In stark contrast with the mainstream perspective, where consumption is assumed to be independent and sovereign to the individual, this "Keeping up with the Joneses" process means that consumption behaviour is not sovereign and is instead dependent on what others do and think.[3] To have a feeling of self-worth a person needs to display their relative wealth via the consumption of expensive goods and excessive amounts of leisure, which is dependent on their relative income. However, much of the economics literature on wellbeing, such as the work of Layard (2011), ignores the depth and importance of relative poverty as a key antecedent to poor mental health, which is critical if low levels of wellbeing are related to social positioning through the "Keeping up with the Joneses" effect.

Veblen (1899) argued that people need to consume conspicuously to highlight their wealth to others, enforce their relative position in society, and climb the social rank. Underpinning these issues are two distinct and

[3] One implication of conspicuous consumption is that more goods are produced than are required for subsistence, which generates unnecessary pollution and uses greater amounts of energy. Raworth's (2018) Doughnut Economics perspective misses the point that the ecological ceiling and social foundations of the doughnut are not the primary focus for individuals within society, and hence the importance of social positioning makes it difficult to believe that society will reduce consumption levels to achieve sustainability.

related activities that Veblen observed: to fit in to obtain the feeling of normalcy and security, and to outdo peers to obtain the feeling of being special. In other words, people seek self-esteem by creating an affiliation with others and by engaging in invidious comparisons towards others. Relative performance is known to be important for an individual's subjective wellbeing (Stutzer, 2004).

People make behavioural choices based on both monetary incentives and their identity (Akerlof & Kranton, 2010): holding monetary incentives constant, people avoid actions that conflict with their concept of the Self. Previous research from anthropological economics (e.g., Douglas & Isherwood, 1996) saw identity as both int*ra*-individual (i.e. psychology) and int*er*-individual (i.e. social), and hence identity is linked to both commitment (Davis, 2004) [i.e., nurturing] and socially influenced preferences (Benjamin et al., 2010). Part of a person's identity is associated with the care they give to others, and van Lange et al. (1997) define a willingness to sacrifice as "foregoing one's immediate self-interest to promote the wellbeing of the partner or relationship."

Marketing and advertisement can manipulate people and channel the projection of an aspirational Self-image into consumer products that promise the attainment of the person that people want to be. A fancy designer suit, designer shoes, overt and "bling" jewellery, and ostentatious watches are symbols of achievement, pride, and strength that allow the consumer to escape from the knowledge and feeling of being inferior. These tastes are displayed as goods that relate to our relative position within society and are referred to as positional goods (Hirsch, 1976). Driving a Ford may suggest that a person is from the middle class, whereas a person who drives a Rolls Royce car may be from an upper class. The frame of reference is also important. Marx (1847) famously emphasised that: "A house may be large or small; as long as the neighbouring houses are likewise small, it satisfies all social requirement for a residence. But let there arise next to the little house a palace, and the little house shrinks to a hut. The little house now makes it clear that its inmate has no social position at all to maintain, or but a very insignificant one; and however high it might shoot up in the course of civilisation, if the neighbouring palace rises in equal or even in greater measure, the occupant of the relatively little house will always find himself more uncomfortable, more dissatisfied, more cramped within his four walls."

The institutional economics perspective is strongly aligned with the social model of mental health (see Beresford, 2002; Mulvany, 2000), for this economics perspective recognises the importance of an individual position relative to those with whom they have stronger ties. Economists preaching this perspective recognise that an individual's non-sovereign tastes, relative positioning, and wellbeing are related to their previous behaviours that evolve with time, practice, knowledge, attention, and understanding of their own experiences and the observable practices and experiences of others, including work colleagues and social influencers.

The self-interested utility maximising perspective treats all individuals as unconnected and as though each person were drawn from a random sample, albeit sometimes with clustering. Figure 9.1 reflects connections that relate to people's behaviours, social positioning, status, confidence, and self-worth, some of which are hierarchical, as shown in this figure, or sometimes cross-classified or even random, which is consistent with Granovetter (1985). If the social model of health and wellbeing has weight, as shown by numerous qualitative studies, then economists must analyse wellbeing data with the inclusion of institutional structures, social networks, social positioning, relative income and poverty, and factors related to social status. Any economic analyses of wellbeing that do not explicitly and completely integrate these social factors into a resource allocation model will invariably lead to predictions that are not accurate reflections of the determinants of wellbeing, thereby rendering the results not even wrong (Woit, 2006).

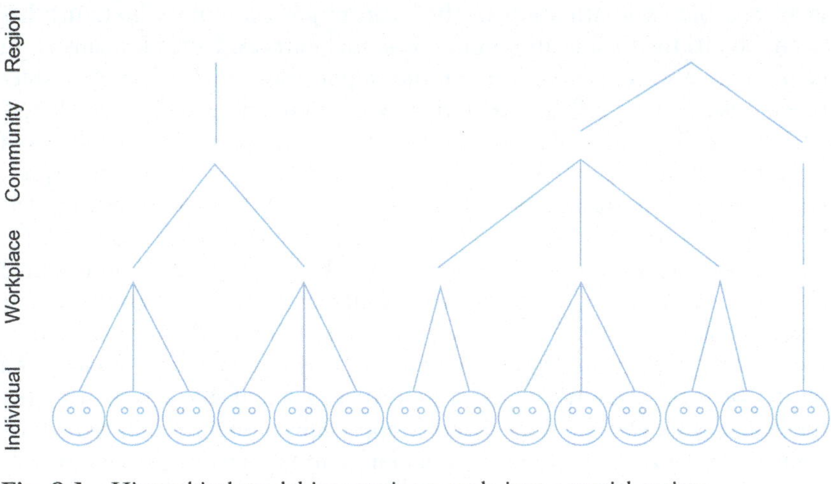

Fig. 9.1 Hierarchical model integrating people into a spatial society

Figure 9.1 also highlights the complexity of acquiring appropriate data to use in econometric models. Social networks can be hierarchical, unstable, and dynamically changing while others are cross-classified, stable, and static. Given the idiosyncratic experiential history and memory of individuals, each layer may have different levels of evolutionary importance for different individuals. Simply including person-specific dummy variables or clustering in an econometric model will not capture the heterogeneity that is so crucial in understanding wellbeing.

Given their significance, it is entirely inappropriate to assume away the importance of social structures on individual health outcomes. When statistical tests are employed to identify whether there is evidence to disprove a null hypothesis—such as there is no relationship between wellbeing and social forces—the strength of that test relies inevitably on the collected data. The existence of partially suitable data, such as geographical area dummies, does not validate the use of models that prove a null hypothesis cannot be rejected. Rather than impairment-focused bio-medical models that adopt mechanistic explanations of causation, a holistic and sociologically informed model of health and wellbeing is both useful and necessary in informing resource-level decision making.

Institutional economists do not necessarily see age, ethnicity, gender, education, employment, income, housing, living arrangement, personal status, social networks, frequency and length of hospitalisation, intervention methods, criminal justice history, and physical ill-health as objective variables, but little attention in the mainstream economics literature has been paid to their subjective and experiential natures. Instead, mainstream economics approaches centre on the supposition that these "variables reflect risk factors related to having and maintaining a mental illness" (Ramon, 2007) and while many of these factors are part of many models of poverty (see Bywaters & McLeod, 1996) the "concepts of poverty, exploitation and power are notably absent" (Ramon, 2007) from social models of health. Likewise, although victimisation, personal attacks, and racial discrimination are specifically mentioned as risk factors for common mental disorders in ethnic minority groups (Meltzer et al., 2004, p. 207) they remain excluded from mainstream economic frameworks. Consequently, the process of exclusion is side-lined in exchange for the development of a model of social causation. The search for cause and effect is akin to the medical model, which is a mechanistic approach to understanding the relationship between mental health, wellbeing, and a person's social environment that restricts our ability to make informed policy decisions.

Are We Moving in the Right Direction?

Econometricians are beginning to estimate the influence of social issues on wellbeing, but their results are arguably not decisive enough to reliably inform resource allocation decisions. An emerging approach compares organisational contexts with levels and changes in wellbeing and/or productivity, such as Torrisi (2013), but of course, the workplace is only one social institution that may correlate with wellbeing, with the household, family, and community being three obvious alternative social institutions that require full and proper integration into any effective wellbeing model. Multi-layered contextualised social factors are well known to be vitally important determinants of wellbeing in the sociology and psychological literatures, and although there are multilevel econometric approaches that could reveal the statistical importance of these issues it remains dependent on the availability of data that reflect the multitude of contemporary and historically relevant interpersonal connections.

Some of the most recognised research on wellbeing in the economics literature, such as Bryson and MacKerron (2017), do provide interesting applications of models that explore wellbeing and/or satisfaction related to the workplace. Studies investigating the connections between organisational culture, wellbeing, and productivity do reveal that the working environment matters. Although these contributions justify the allocation of wellbeing enhancing resources to companies, questions remain whether the sensitivity of workers to their work environment is moderated by the social ties of friends and family.

There are links between wellbeing and labour market performance (e.g. Judge et al., 2001; Lyubmirsky et al., 2005) and evidence of a positive association between wellbeing and worker productivity (Oswald et al., 2014), but evidence based on observations within workplaces is problematic for several reasons. First, just because there is a lack of published economic evidence detailing the importance of other social institutions (such as family ties) does not mean that they are not important for worker wellbeing, and nor does it mean that studies have not sought to identify such a causal link. The lack of attention towards the influence of workplace employee relations (including flexible working arrangements, job design, and inclusive workplace cultures) on wellbeing after controlling for the influence of family and friends misses an avenue for policy attention. Second, economic evaluations of the effect of a treatment on firms need to factor in two crucial considerations: the time delay and cumulative effects of the treatment, and the internal and external contextual effects that may limit the impacts of those treatments. Additional complications exist here as the intensity of any time-delayed effect will be affected by an individual's prior experiences, and hence prior experiences could influence the strength and duration of any effects of policy implementations on wellbeing.

Recommendations and Conclusions

Governments need an evidence base to steer policy formation, but this is attracting incomplete and underinformed empirical findings. Either we fund expensive large-scale longitudinal studies to collect data encompassing the myriad of possible socially relevant contextually embedded social factors that underpin fluctuations in wellbeing, or we recognise and accept that small scale deep qualitative studies are going to be more meaningful than the results of existing number-rich-theory-poor economic models. We propose the following as critical for the transformation of the social system.

First, there needs to be a methodological shift in the mainstream economics discipline. The economics discipline is likely to dominate resource allocation decisions for a long time to come, which emphasises (1) the need for broader education for economics students at universities (including teachings on Adam Smith's *Theory of Moral Sentiments*), (2) deeper training on the limitations of econometrics, and (3) wider training research methodologies; doing so should explicitly appreciate other-regarding.

Second, the dominance of economists in making resource allocation decisions is natural given that the economics discipline provides a substantial amount of training in resource allocation decision making. It is important to recognise the very high regard for the economists in most government departments, partly because many of the individuals that work in government departments also have some training in economics. However, perhaps partly due to their narrow economics training, few government officials appear to appreciate the breadth and wealth of knowledge of institutional economists, and the mainstream economists currently advising on policy would not recommend them because they disagree. Many institutional economists are available to provide economic advice on wellbeing related resource allocation decisions and they would do so without relying on the discipline's mainstream flawed implicit ontological, epistemological, and axiological foundations.

Third, resource allocation decisions need to be formulated with input from all relevant disciplines. We advocate for multidisciplinarity, free from exclusion and veils of superiority, and that does not require acceptance of an ontology. The dominance of mainstream economics in guiding government resource policies impedes the free flow of interdisciplinary knowledge and limits contributions from other social sciences. After all, we know from Toffolutti and Suhrcke (2019) and Mills (2018) that economic policies can kill.

Fourth, and not least, there needs to be a greater appreciation that it is not possible to estimate effective models without appropriate data. Omitted variable bias is widespread in economic analyses of wellbeing; including proxy variables that are partially correlated with a social factor will inevitably incur measurement error and lead to spuriously accurate results. The omission of social factors in a social wellbeing model creates results that are not even wrong. The lack of awareness of the limitations of econometric models for policymaking is a serious oversight, and we need to recognise that social policies should be based on social awareness, not mathematical elegance. Embracing insights from across the social research disciplines would surpass the power of positivist economic analysis.

Mastering a seismic change that promotes a socialised model of wellbeing requires the devotion of resources over a prolonged period of time, although power and incentive mechanisms are against this necessary development (Webber & Pattinson, 2022). The adoption of a multidisciplinary approach that encompasses broader institutional economics thought would embrace the importance of social networks, sociological, and spatial factors and permit a more appropriate allocation of resources that could more effectively transform the social system and enhance wellbeing for all.

References

Akerlof, G. A., & Kranton, R. E. (2010). *Identity economics*. Princeton University Press.

Barnes, M., Davis, A., & Tew, J. (2000). Valuing experience: users' experiences of compulsion under the Mental Health Act 1983. *Mental Health Review Journal, 3*(5), 11–14.

Benjamin, D. J., Choi, J. J., & Strickland, A. J. (2010). Social identify and preferences. *American Economic Review, 100*(4), 1913–1928.

Beresford, P. (2002). Thinking about 'mental health': Towards a social model. *Journal of Mental Health, 11*(6), 581–584.

Black, C. (2008). *Working for a healthier tomorrow*. Department of Work and Pensions, UK Government.

Bryson, A., & MacKerron, G. (2017). Are you happy while you work? *Economic Journal, 127*, 106–125.

Bywaters, P., & McLeod, E. (1996). *Working for equality in health*. Open University Press.

Davis, J. (2004, May). Identity and Commitment: Sen's Conception of the Individual. *Workshop on Rationality and Commitment*, 13–15. University of St Gallen.

Douglas, M., & Isherwood, B. (1996). *The world of goods: Towards an anthropology of consumption*. Routledge.

Dugger, W. M. (1989). Instituted process and enabling myth: The two faces of the market. *Journal of Economic Issues, 23*(2), 607–615.

Granovetter, M. (1985). Economic action and social structure: The problem of embeddedness. *American Journal of Sociology, 91*(Nov), 481–510.

Harvey, J. (2016). *Contending perspectives in economics: a guide to contemporary schools of thought, Edward Elgar*. Cheltenham.

Hirsch, F. (1976). *Social limits to growth*. Harvard University Press.

Hume, D. (1739-40). *A treatise of human nature*. Clarendon Press.

Judge, T., Thoresen, C., Bono, J., & Pattern, G. (2001). The job satisfaction-job performance relationship: A qualitative and quantitative review. *Psychological Bulletin, 1277*(3), 376–407.

Layard, R. (2011). *Happiness: Lessons from a new science*. Penguin Books.

Lyubmirsky, S., King, L., & Diener, E. (2005). The benefits of frequent positive affect: Does happiness lead to success? *Psychological Bulletin, 131*(6), 803–855.

Marx, K. (1847). *Wage labour and capital*. Pamphlet.

Meltzer, D., Fryers, T., & Jenkins, R. (2004). *Social inequalities and the distribution of the common mental disorders*. Psychology Press.

Mills, C. (2018). 'Dead people don't claim': A psychopolitical autopsy of UK austerity suicides. *Critical Social Policy, 38*(2), 302–322.

Mulvany, J. (2000). Disability, impairment or illness? The relevance of the social model of disability to the study of mental disorder. *Sociology of Health & Illness, 22*(5), 582–601.

Myddelton, D. R. (2009). *'Spurious accuracy,' margins of error in accountancy*. Palgrave Macmillan.

Oswald, A., Pronto, E., & Sgroi, D. (2014). Happiness and productivity. *Journal of Labor Economics, 33*(4), 789–822.

Pilgrim, D., & Rogers, A. (2003). Mental disorder and violence: An empirical picture in context. *Journal of Mental Health, 12*(1), 7–18.

Ramon, S. (2007). Inequality in mental health: The relevance of current research and understanding to potentially effective social work responses. *Radical Psychology, 6*(1), 1–23.

Raworth, K. (2018). *Doughnut economics: Seven ways to think like a 21st century economist.* Random House Business.

Smith, A. (1759). *Theory of moral sentiments.* Cambridge University Press.

Smith, A. (1959). *The theory of moral sentiments.* Penguin

Stutzer, A. (2004). The role of income aspirations in individual happiness. *Journal of Economic Behaviour and Organisation, 54*, 89–109.

Toffolutti, V., & Suhrcke, M. (2019). Does austerity really kill? *Economics and Human Biology, 33*, 211–223.

Torrisi, B. (2013). Academic productivity correlated with well-being at work. *Scientometrics, 94*(2), 801–815.

van Lange, P. A., Agnew, C. R., Harinck, F., & Steemers, G. E. (1997). From game theory to real life: How social value orientation affects willingness to sacrifice in ongoing close relationships. *Journal of Personality and Social Psychology, 73*(6), 1330.

Veblen, T. (1899). *The theory of the leisure class: Economic study of institutions.* Random House.

Webber, D. J., & Pattinson, S. (2022). Social pressures to conform to a given set of research methods. In P. Sandiford & S. Schuhrer (Eds.), *Exploring Research Methodology and Research Design*. Edward Elgar.

Woit, P. (2006). *Not even wrong: The failure of string theory and the search to unify the laws of physics.* Basic Books.

CHAPTER 10

Social and Material Foundations of Wellbeing: Beyond the Neoliberal Model of Development

David Neilson

Abstract By unleashing capitalist social relations on a planet-wide scale, the "neoliberal model of development" is destroying the possibility of what the young Marx (Economic and philosophical manuscripts. In K. Marx. *Early writings*. Penguin in association with New Left Review, 1975) called humanity's "species [well]being" defined as material security, inclusive solidarity and creative autonomy. The first part of the chapter, in opposition to Althusser's anti-humanist dismissal of the early work, builds on young Marx's alienation-based account of how capitalist production relations undermine humanity's wellbeing. The second part demonstrates how capitalism's contemporary neoliberal form further undermines the foundations of human wellbeing. Building critically beyond Marx's nineteenth-century critique, this chapter argues that the competitive, exclusive, precarious, individualistic, ecologically unsustainable, and

D. Neilson (✉)
Sociology, The University of Waikato, Hamilton, New Zealand
e-mail: david.neilson@waikato.ac.nz

A. H. Kemp, D. J. Edwards (eds.), *Broadening the Scope of Wellbeing Science*, https://doi.org/10.1007/978-3-031-18329-4_10

COVID-19 -ridden contemporary neoliberal-led capitalist world is undermining the social and material foundations of humanity's wellbeing considered as "planetary wellbeing" (Antó et al., Sustainability 13:3372, 2021). The concluding discussion sketches elements of a democratic socialist alternative to both the neoliberal model of development and to authoritarian socialism that can deliver planetary wellbeing.

Keywords Marx • Alienation • Human nature • Capitalism • Neoliberal model of development

Introduction

Beyond the usual critiques of neoliberalism that emphasise ideology and are "methodologically nationalist", this chapter brings in its "model of development" (Neilson, 2020a). This concept refers here to the widespread adoption of a national regulatory template that in aggregate creates a trans-national regulatory environment that defines the limits, possibilities and priorities of national state agency. The neoliberal model of development, grounded in the widespread adoption of the neoliberal national template, has resulted in the contemporary global terrain of free market competition on which capital is free to move. In turn, countries must chase this globally mobile capital to be viable, which thereby subordinates them to capital's priorities and transforms them into "competition states" that must contest with each other for scarce capital. The logic of global market competition between capitalist firms reflected in the competition between countries has generated a deeply unstable and alienating form of human civilisation's development.

The neoliberal model of development has led to planetary "malbeing". Specifically, it has unleashed the neoliberal-led contemporary global market competition form of contemporary capitalism that is driving rapid ecological destruction, giving rise to regressive forms of competitive nationalism and individualism, intensifying material precarity and uncertainty, and is the key driver of the global COVID-19 pandemic (Neilson, 2021b). Transforming these material and social foundations of malbeing into those of wellbeing defines the calculated goal of an alternative democratic socialist model of development that is proposed later in in this chapter. To begin with, building on Marx's theory of humanity's estrangement from its authentic core needs, this chapter discusses the problematic concept of wellbeing.

Alienation and Wellbeing

Young Marx (1975) outlines a view of humanity's core authentic needs that when met constitute wellbeing. Like all sentient beings on this planet, Marx argues humanity's wellbeing depends primarily on material security that is contingent on both the state of Nature and, for humanity in particular, on the social and technical form of the mode of production. Specifically, he builds his conception of humanity's organic wellbeing defined as material security, and linked with inclusive solidarity and creative autonomy from his critique of the alienating consequences of capitalism. Consistent with Marx's discourse, Giddens (1991) refers to "ontological security", i.e., a materially secure daily life pattern facilitating a feeling of ease that is basic to wellbeing. Inversely, "existential anxiety", manifesting as a physical dread in the pit of one's stomach, arises when daily life is precarious because its material conditions are scarce, uncertain, and possession is contested.

Material security is always and everywhere the primary condition of humanity's wellbeing. However, under capitalism the young Marx observes that the labouring population's circumstances are characterised by material precarity because without independent means of life, they are existentially dependent on those who own these means. Without a viable alternative, the labouring population is compelled to enter into the precarious terms of an exploitative wage relation with capital that is characterised by uncertain employment and minimal payment. Further, as expressed by Marx and Engels in the 1845 *German Ideology*, deeply volatile and unstable processes beyond the labouring population's control determine its basic life situation.

From the premise of wage dependent precarity, Marx (1975) builds his theory of alienation. By entering into the wage relation, labour cedes to capital ownership and control of the conditions, process, and results of its fundamental life activity. In consequence, rather than directly expressing life, the labour activity is reduced to being a means of existence that "mortifies the flesh and destroys the mind", all the while growing capital's wealth and power alongside labour's poverty and powerlessness. Inversely, as crystallised in the 1867 Bees and Architect's tale of *Capital* Vol. 1 (1976), Marx argues that what distinguishes the productive life of humans from bees and spiders is that humans are creative producers who imagine before practically producing products to meet their needs. What differentiates authentic human nature from bees and spiders is that the former's essential nature is consciously and experimentally creative. Therefore, though results tend to fall short of what is originally imagined, over time humanity can accumulate knowledge from its praxis. In sum, according to Marx, humans are naturally creative producers but under capitalism this life-affirming activity is turned into a forced and dehumanising means of precarious existence.

Further, the alienated experience of the labour activity undermines humanity's capacity to recognise or embrace the labour activity as a counter-factual imaginary expressing authentic human needs. In turn, Marx infers that this "self-estrangement" undermines human solidarity. Though positing human nature as intrinsically social, that is, wellbeing requires social bonds both for material security and as an end-in-itself; Marx argues that the labouring population's life under capitalism undermines experience and self-recognition of its authentic socially creative nature. This self-estrangement implies mutual estrangement and therefore also estrangement from one's "species-[well] being". The labouring population's solidarity is also undermined by competition and by "commodity fetishism" whereby the exchange relation between things obscures the cooperative basis of the social division of labour (Marx, 1976). However, Marx is loath to recognise how this argument contradicts his optimistic prognosis that labour's circumstances will give rise to an inclusive solidarity. Durkheim faces a similar problem when he realises that the modern division of labour actually does not directly generate "organic solidarity" as he originally argued but, rather, requires the counter force of "moral regulation" in the form of shared values (Durkheim, 1984). Without moral regulation, the complex division of labour of modern market coordinated society can sometimes lead to "anomie" expressed as the breakdown of social solidarity into a lonely mass of socially isolated individuals (Durkheim, 2005). In short, rather than giving rise to solidarity, as they would like, Marx and Durkheim's own analyses of modern capitalism also indicate strong tendencies in the opposite direction.

In contrast to capitalism that estranges humanity from its trans-historical authentic life-affirming nature, Marx's position is that wellbeing depends on a material and social context that affirms and facilitates humanity's core authentic needs defined as creative autonomy, social solidarity, and material security. Durkheim, in addition, specifies a discursive dimension to solidarity. As the inverse of alienation, wellbeing then refers to the situation where material, social, discursive relations, and natural dimensions of human wellbeing are mutually confirming. For Marx, the relationship between these dimensions is complicated because through the history of class-based modes of production, manifest expressions of human nature are the products of specific material, social and ideological circumstances that do not straightforwardly reflect authentic human needs and goals. Marx thus holds a variation of the familiar view that the practical expression of human nature is always the product of the complex interaction between an authentic core and historically specific set of circumstances that are not simply reflective of each other. This perspective raises difficult epistemological and ontological problems.

First, how can one demonstrate what constitutes authentic human nature when it can never be observed because in practice it is always circumstantially overdetermined? Neoclassical Economics avoids this problem by naturalising forms of human behaviour that are assumed to correspond causally with an idealised celebration of market capitalism. Specifically, the market order is posited as arising spontaneously from humanity's selfish, competitive, and lazy consumerist nature. Paradoxically, in neoclassical labour market economics, problems of unemployment, apparently caused by labour's naturally selfish greed and laziness, are countered by encouraging the "work ethic" and tempering sectional self-interest.

According to Marx, mainstream Economics gets the causality the wrong way round. Rather than humanity's core nature generating a natural spontaneous social order; the social structural form and logic of capitalism independent of the labouring population's will and consciousness brings out certain -while frustrating other-aspects of human nature. Under capitalism, where circumstances are characterised by scarcity combined with zero-sum competition and wage dependence, engagement of human nature's life-instinct becomes more about survival than about wellbeing. In essence, competitively insecure circumstances engage humanity's competitive instincts in order to survive. In addition, alienating work and "commodity fetishism" under capitalist relations, lead humanity to look for ways to compensate for alienation and quell resulting feelings of anxiety by buying and consuming new things. Carlisle et al. (2009), following a similar line of analysis, demonstrate how the "science of wellbeing" accepts that humanity's actual social and material circumstances can be at odds with what it needs to flourish rather than just survive.

However, second, in practice human nature always takes the form of a complex interaction with a less than ideal social-material context. That is, manifest forms of human nature express the causal interaction between social-material circumstances and humanity's core nature. Although a strong life-instinct is an invariantly persisting feature of core human nature across different historical epochs, different life-seeking responses to different material-social contexts complexly relate to early evolutionary developments in this trans-historical core. That is, wellbeing is premised fundamentally on a social and material context of life confirming security, that links with solidarity and self-control, and that puts us at ease mentally and physically. However, humanity's deep survival instincts expressed as competitive aggression, regressive individualism, "fight or flight" and submission have primordial evolutionary origins in extreme situations such as deep material precarity, jungle competition, and/or slavery. Thus, although humanity's "malbeing" refers to being ill-at-ease both mentally and physically in a precarious material/social context, embedded in humanity's complexly evolved nature are also a range of natural instincts that have enabled it to survive such contexts.

Leaving aside the question of the complex evolutionary relation between core human nature and its changing social and material contexts, this chapter assumes a complexly multi-sided trans-historically constant core that manifests selectively and differently in response to changing social-material contexts. Thus, for example, while human wellbeing requires social conditions facilitating both solidarity and autonomy, one imagines that as the social form of human existence has become more individualised and less directly communal, so too has humanity's historically manifest nature. More complexly, people adapt their core needs in response to alienating circumstances. They seek pleasure in the purchase and consumption of things as a form of compensation for the alienation caused by its fetishised form. In Lacanian terms, this adaption is a manipulation fostered by advertising. That is, advertising presents the myth of an absent "jouissance" that will be fulfilled by the purchase of the advertised product, but that actually only gives a temporary fantastical fleeting pleasure (Stavrakakis, 2000). Giddens (1990) argues, in the same general direction, that mental states associated with alienation and unease can find pleasurable expression. For example, though under globalisation existential anxiety abounds due to the former's dynamic uncertainty, the later can manifest as excitement and exhilaration. More pessimistically, building on Layton's (2010) association of Freud's concept of disavowal and neurotic ways of adaptively responding to alienating circumstances under neoliberalism, Neilson (2015) discusses different forms of denial, associated with different class circumstances, which function to quell and divert existential anxiety.

Differing material and social contexts selectively activate different combinations and adaptions of humanity's complex multi-sided core nature that in addition varies in its exact composition across the human population. In historical practice, contradictions between authentic core needs and social material contexts generate creatively adaptive manifestations. The flourishing of humanity's authentic nature therefore does not simply equate with wellbeing defined as security, solidarity and autonomy. Rather, complex human nature schematically includes, and thus the expression of humanity's core nature involves, not only needs associated with wellbeing but also those relating to more aggressive, competitive and selfish instincts that have facilitated survival in conditions of material insecurity and coercive subordination. Relatedly, people may become attached to their "alienated self-hoods". Moreover, we creatively adapt to alienating circumstances making it possible that despite everything "we might just get what we need". The changing material and social conditions of life that under capitalism involve competitive precarity and unsatisfying work also give rise to the pursuit of creative compensations for the absence of the conditions of wellbeing, and/or exciting less than harmful expressions of competitive survival instincts. Nonetheless, a material and social context facilitating ease and positive wellbeing is still treated here as humanity's primary goal, and the other sides of human nature though needing outlet and expression are best served by recognising and adapting them to constructive expressions compatible with the primary positive pole.

Transcending the Precarity of Neoliberal Globalisation: Creating Alternative Material and Social Foundations for Planetary Wellbeing

Contrary to prevailing readings of Marx, this chapter has built on the early analysis of alienation understood as the capitalist negation of wellbeing. Human nature is multi-sided and interactively adaptive to different environments, yet there are still basic differences between life-destroying alienation and life-affirming wellbeing. However, following Gautama, wellbeing has absolute and relative dimensions. He understood that physical and mental suffering inevitably, i.e., absolutely, ending in death ("dukkha") is intrinsic to the human condition: "we are born, we grow old, we get sick, we die". In an existential reading, humanity's self-reflexive-knowledge of dukkha manifests as existential anxiety. In the "four noble truths", Gautama refers to unhealthy ways humanity quells this anxiety by forms of denial, or "grasping and craving" (see Batchelor, 2012). He recommends a healthy life of calm meditative acceptance of the absolute organic limits of life combined with being mindfully present in a comfortable social and material environment. Following this path, people are able to ease, relatively, existential anxiety and attain a feeling of oneness with each other and the universe. Resonating both with Marx and Gautama's formulation, Giddens' (1991) concept of "ontological security" refers to a (relative) feeling of ease generated by the reassuring continuity of life-confirming everyday circumstances and experience centrally including material security, trust and solidarity, and control over our lives. In sum, *relative* wellbeing is grounded in, though not the automatic effect of, a material and social environment that facilitates universal friendship, creative cooperation, and material security.

What is usually less visible is "planetary wellbeing" understood as involving the bigger, deeper and faraway social and material conditions of everyday life. Moreover, even when such conditions begin to directly express themselves as a disturbing anti-wellbeing break or catastrophic discontinuity in the everyday pattern we may, argues Freud (1927) and resonating with Gautama's concept of "grasping and craving", "disavow" them in order to avoid the anxiety that occurs when we confront them. Building the foundations of an optimistic alternative to fatalistic confrontations with an existing catastrophic planetary reality offers a way forward beyond denial. Building on my own works, this environment is deliberately sought by the design and deployment of a post-neoliberal, democratically cosmopolitan, cooperative, and ecologically sustainable model of development (Neilson, 2020a, 2020b, 2021a, 2021b).

Descent Towards the Abyss

Marx's 1844 (1975) analysis of invariantly present social and material conditions of alienation under capitalism and his mature 1867 (1976) analysis in *Capital* Vol. 1 of their deepening with capitalism's maturity are epochal in their significance. Borne out by contemporary automation of both mental and manual work is Marx's analysis that capitalism's deep logic of "expanded reproduction" generates a continuing rise in the "organic composition of capital" (OCC), i.e., the ratio of "fixed capital", or dead labour congealed in the means of production, to "variable capital" or living labour. To the present, capital's exploitation of living labour continues to accumulate the dead labour of "self-acting machinery" that subordinates and then subsequently replaces living labour. Though practically grounded in workers' living labour, this process increasingly negates the labouring population's wellbeing because it takes the form of capital's expanding alien, controlling, and destructive power. Further, contemporary capitalism's interacting crisis tendencies are centrally related to its key downstream effect of a growing "relative surplus population", that in turn are driving the labouring population's deepening precarity and alienation today.

Mechanisation is central to capital's dehumanisation of the industrial labour process that turns workers into the appendages of the unvarying continuous movement of the machine system (see Neilson, 2007). The rising OCC has also fragmented the labouring population's class structure into ***knowledge workers***, who drive the technological subordination and redundancy of the "active army" of the industrial ***working class***, who decline as a numerical proportion of the labouring population, leading to the growing ranks of a deeply insecure and impoverished ***relative surplus population*** (Neilson, 2018; Neilson & Enright, 2022). First, this process leads to deepening uncertainty around the conditions of material existence by driving wages down while growing the "Reserve Army" of those temporarily (un)employed. Second, labour's oversupply that drives wages down leads to the extending of working time, the increasing of capital's power, and inversely, labour's growing precarity because employers can easily dismiss workers when others are waiting in line (Neilson, 2007). Relatedly, third, increasing competition for scarce employment amongst the labouring population implied by a growing relative surplus population undermines trust and solidarity and unleashes competitive survivalism.

This process is also at the core of capitalism's recurring economic crises driven by imbalances between production, that grows with competition-driven mechanisation, and consumption that declines because the relative surplus population grows. Certainly, there are tendencies that counter the relative surplus population tendency including the employment of more knowledge workers, the retention of production workers making more and different products, and increasing non-productive service work. However, the indefinite increase in material output, implied by employing more workers who with advanced technology can produce ever-greater volumes of material products, is at the core of the destructive consumption of Nature that is existentially confronting humanity today as a deepening multi-sided ecological crisis.

Crystallising in the mid-1980s under the Washington Consensus, the contemporary "neoliberal model of development" has reversed the Keynes-led progressive model that existed amongst the advanced capitalist countries from 1945 through to the late 1970s (Neilson 2020a). Constructed by the more or less widespread adoption of Keynes' national template, this postwar model of development comprised a mutually reinforcing national-trans-national nexus. The widely adopted national template was centrally concerned with the organised management of wage-led domestic demand so that it increased in line with increasing productivity, combined with the limiting of capital's international mobility. Counteracting international competition by reducing the motivation and capacity for capital flight was central to the template's viability understood as a nation-centred accumulation strategy. In direct contrast, the neoliberal model of development has pushed all countries to open themselves up, both within and across territorial borders, to capital's free movement. In aggregate global effect, this neoliberal model of development has intensified humanity's planetary "malbeing" by unleashing capitalism's demand depressing competitive logic and profit imperative, including that between countries and their workforces, and extending it globally.

In addition, beyond *Capital* Vol. 1's prognostic account, capitalism has unfolded unevenly over time and across different geo-political territories (Neilson, 2020a). Under the neoliberal model of development, this uneven process has been intensified not only by "locational competition" between countries but also by the simultaneous activation of two distinct stages of the relative surplus population tendency, i.e., dispossession of the peasantry in Global South countries associated with industrial capitalism at its early stages, and increasing automation leading to industrial workers' redundancy that occurs with industrial capitalism's mature stages (Neilson, 2020a). In turn, these viciously interacting factors have made locational competition between countries zero-sum because labour's oversupply is capital's scarcity. At the same time, this competition is intensifying the wage race to the bottom that is at the core of recurring international economic crises, while increasing production is simultaneously destabilising the planet's ecology at an alarmingly increasing rate (Neilson, 2021b). Further, inscribed in this zero-sum logic of aggressive locational competition, and running contrary to neoliberalism's ideological aspiration, is the growth of "regressive nationalism" (Neilson, 2021a). That is, zero-sum locational competition that threatens national economic viability including local employment and local ways of life feeds anxiety, fear, and mistrust of the "other" who are located within and beyond national borders. Such psychological responses, in turn, provide the perfect emotional base for xenophobic and racist forms of regressive nationalism that can release this anxiety as anger. Finally, overlaying these forms of the global negation of humanity's wellbeing is the global pandemic of COVID-19. Capitalism's march into the wilderness and neoliberal-led economic globalisation of people through tourism and production through global supply chains on which countries have become dependent has spread and embedded the virus across the entire planet (Neilson, 2021b).

In sum, beyond even Marx's nineteenth-century critical prognosis, the neoliberal model of development has unleashed a dramatically intensifying and extending planet-wide expression of capitalism's deep crisis tendencies. The contemporary vicious cycle of interacting crises centrally includes the deep and volatile instability of global economic reproduction, an unstable, unequal and competitively alienating class structure, a deeply precarious relative surplus population, destructive individualisation, the rise of neo-fascist forms of regressive nationalism, deepening ecological destabilisation, and a global pandemic. We are descending towards a "planet of slums" (Davis, 2006) comprising diseased and malnourished people surplus to capitalism's productive requirements struggling in contestation with each other for scraps of work and food for their very survival, that is being intensified by the broader context of economic and political instability combined with deepening Natural scarcity, destruction and destabilisation. In sum, the contemporary neoliberal-led capitalist world is dramatically moving in the opposite direction of a sustainably secure and a cooperative and inclusive social relational structure that is the pre-condition of humanity's wellbeing.

Contra orthodox Marxist readings, the conditions of a democratic socialist world of universal material plenitude, individually confirming social solidarity and free creativity are not inscribed in capitalism's dynamic of extended reproduction. Rather, seemingly beyond our ability to reverse or control, the logic of contemporary capitalism is driving the rapid descent of human civilisation towards material destruction, social breakdown and competitive survivalism. Rather than towards the social and material conditions of wellbeing, the current crisis of global capitalism is leading, albeit unevenly across the world's population, towards conditions of deep precarity and alienation (Neilson, 2021b).

Emerging from the Wreckage: So What … Is To Be Done

An alternative model of development is urgently needed that can reverse the catastrophic effects of the neoliberal model and create the deeper material and social foundations of planetary wellbeing. The proposed progressive alternative to the neoliberal model specifically aims to overcome the blind forces of global market competition and private profit logic that are destroying the planet's ecosystem and making humanity's material civilisation precarious, competitive, and beyond its control. The alternative proposed here is premised on the renovation of the United Nations, politically as cosmopolitan democratisation, and ideologically via the imaginary of delivering universal material security by deploying a cooperative strategy to facilitate ecologically sustainable modes of local accumulation. This is a framework of global solidarity and the prioritising of an associated socialised intensification of virtual globalisation dedicated to democratic communication that prioritises localised self-sufficiency of material production (see Neilson, 2021b). In sum, the proposed model of development seeks to cooperatively facilitate virtual global integration with local self-sufficient autonomy in order to achieve complementarity between the primary goal of universal material security centrally including ecological sustainability, and, solidarity and autonomy. Local material accumulation, within enhanced virtual globalisation, or "glocalisation", is the key to reversing the multi-pronged crisis logic of the neoliberal model of development including economic instability, international rivalry, western domination, regressive nationalism, ecological degeneration, and global pandemics (Neilson, 2020b, 2021a, 2021b; Neilson & Enright, 2022).

Rather than globalised competition and production, the proposed alternative model would centre on facilitating each country's capacity to develop locally on its own terms (Neilson, 2021b). The key would be a flexible national regulatory template. Although having universal conditions regarding wellbeing and ecological standards, this template would be flexibly sensitive to, and grounded in, each local situation and culture and would assert the power of nation states over capital to decide what they import and how they democratically regulate local production and consumption. Local development would be assisted via institutions of regional cooperation, and a United Nations "World Knowledge Bank" that would acquire and then socialise cosmopolitan knowledge combined with providing practical resources and expertise that involve the enlistment of capitalist firms (Neilson, 2020b). In order to achieve universally viable local development, the global capitalist project of pervasive market competition and private ownership would be countered and capitalist firms enlisted to serve the goals of conscious cooperative nationalism.

Concluding Comments: "So What"

The current neoliberal mode of capitalism that negates wellbeing understood as security, solidarity, and autonomy contrasts with a brief outline of an alternative model of development consciously designed to facilitate humanity's wellbeing. The argument is nuanced because it recognises that the negative dimension of wellbeing, or "malbeing", that is associated with scarcity, precarity, and competitive individualisation, also locates embedded elements of core human nature that require expressive outlets. While recognising security, solidarity and autonomy as primary, this chapter has also argued that humanity needs to express other dimensions of its core nature, especially those that relate to individualistic competitive instincts. Relatedly, while material/social security is a relatively straightforward primary principle, the relationship to, and between, freedom and solidarity require much more careful articulation if they are to be compatibly confirming of each other. Integrally, the way forward towards the social and economic foundations of humanity's wellbeing is also not simply one-sided. Practically, it is not a matter of, absolutely, replacing competitive precarity with cooperative material security, or for that matter global-scale production with that on a local scale. Rather, the goal is to subordinate the former to the latter which is expressed in constructive, relative, and affirming forms.

References

Antó, J. M., Martí, J. L., Casals, J., Bou-Habib, P., Casal, P., Fleurbaey, M., Frumkin, H., Jiménez-Morales, M., Jordana, J., Lancelotti, C., et al. (2021). The planetary wellbeing initiative: Pursuing the sustainable development goals in higher education. *Sustainability, 13*, 3372.

Batchelor, S. (2012). A secular Buddhism. *Journal of Global Buddhism, 13*, 87–107.

Carlisle, S., Henderson, G., & Hanlon, P. (2009). 'Wellbeing': A collateral casualty of modernity? *Social Science and Medicine, 69*, 1556–1560.

Davis, M. (2006). *Planet of slums* (Vol. 23, p. 6). Verso Books.

Durkheim, E. (1984). *The division of labour in society*. Macmillan.

Durkheim, E. (2005). *Suicide: A study in sociology*. Routledge.
Freud, S. (1927). Fetishism. In J. Strachey (Ed.), *The Standard Edition of the Complete Psychological Works of Sigmund Freud: Vol. 21* (pp. 152–159). Hogarth Press. [translated].
Giddens, A. (1990). *The consequences of modernity*. Polity.
Giddens, A. (1991). *Modernity and self-identity: Self and society in the late modern age*. Polity.
Layton, L. (2010). Irrational exuberance: Neoliberal subjectivity and the perversion of truth. *Subjectivity, 3*(3), 303–322.
Marx, K. (1975). Economic and philosophical manuscripts. In K. Marx (Ed.), *Early Writings*. Penguin in association with New Left Review.
Marx, K. (1976). *Capital: Vol. 1*. Penguin in association with New Left Books.
Neilson, D. (2007). Formal and real subordination and the contemporary proletariat: Recoupling Marxist class theory and labour process analysis. *Capital and Class, 91*, 90–124.
Neilson, D. (2015). Class, precarity, and anxiety under neoliberal global capitalism: From denial to resistance. *Theory and Psychology, 25*(2), 185–201.
Neilson, D. (2018). In-itself for-itself: Towards second generation neo-Marxist class theory. *Capital & Class, 42*(2), 273–296.
Neilson, D. (2020a). Bringing in the 'neoliberal model of development'. *Capital & Class, 44*(1), 85–108.
Neilson, D. (2020b). The democratic socialisation of knowledge: Integral to an alternative to the neoliberal model of development. In M. Peters, T. Besley, P. Jandrić, & X. Zhu (Eds.), *Knowledge Socialism. The Rise of Peer Production: Collegiality, Collaboration, and Collective Intelligence*. Springer.
Neilson, D. (2021a). Beyond regressive nationalism and the neoliberal model of development. *Review of Radical Political Economics, 53*(1), 17–34.
Neilson, D. (2021b). Reversing the catastrophe of neoliberal led-capitalism in the time of coronavirus: Towards a democratic socialist alternative. *Capital & Class, 45*(2), 191–214.
Neilson, D., & Enright, N. (2022). From dead information to a living knowledge ecology. In M. A. Peters, P. Jandric, & S. Hayes (Eds.), *Bioinformational Philosophy and Postdigital Knowledge Ecologies*. Springer.
Stavrakakis, Y. (2000). On the critique of advertising discourse: A Lacanian view. *Third Text, 14*(51), 85–90.

CHAPTER 11

Discussion: Broadening the Scope of Wellbeing Science

Andrew H. Kemp and Darren J. Edwards

Abstract This chapter reviews some of the key ideas emerging from the heterogeneous discipline of wellbeing science and presented in this book by individual contributors, encompassing perspectives from psychological scientists, public health professionals, and economic and political theorists. Contributions document a wide range of issues including the expansion of positive psychology to existential issues, including hardship, suffering and levels of scale that usually lie beyond the remit of the field; the application of positive psychotherapy and acceptance and commitment therapy at multiple levels, facilitating opportunities that would not have

A. H. Kemp (✉)
School of Psychology, Faculty of Medicine, Health and Life Science, Swansea University, Swansea, Wales, UK
e-mail: a.h.kemp@swansea.ac.uk

D. J. Edwards
Department of Public Health, School of Health and Social Care, Faculty of Medicine, Health and Life Science, Swansea University, Swansea, Wales, UK
e-mail: D.J.Edwards@Swansea.ac.uk

A. H. Kemp, D. J. Edwards (eds.), *Broadening the Scope of Wellbeing Science*, https://doi.org/10.1007/978-3-031-18329-4_11

otherwise emerged; developments in ecopsychology, highlighting opportunities for individual and collective action to cultivate connections to nature and nature-based conservation; perspectives from public health, focused on building population wellbeing, contextualised by forward thinking government legislation from an emergent wellbeing economy; reflections on the structural limitations of mainstream economics, impacting on the appropriate allocation of resources for a transformation of society; and a critique of the neoliberal model of development from the perspective of social, material and planetary wellbeing. Perspectives are distinct, although overlapping and complementary, providing a strong foundation for an emerging interdisciplinary and even, transdisciplinary model of wellbeing, encompassing the individual, community and planet.

Keywords Social and natural ecologies • Ecological theory • Interdisciplinarity

The preceding chapters present perspectives on wellbeing from colleagues from a wide-range of research areas including positive (Wong and Lambert), clinical (Gibbs and Edwards), and environmental psychology (Richardson and Corral-Verdugo), public health (Edmonds) as well as economic and political systems (Webber and Neilson). Perspectives are overlapping and we hope that these will help to move the field of wellbeing science forward by emphasising perspectives of wellbeing across research areas that are often focused on in isolation. It is helpful to view this work through a social ecological lens, which places the person in the context of their social and natural ecologies. The model of Bronfenbrenner (Bronfenbrenner, 1977) and its adaptation in wellbeing science (Fisher et al., 2020; Kemp & Fisher, 2022; Lomas, 2015; Lomas et al., 2015) highlight opportunities to explore complex interactions across multiple levels of scale, including the individual, the microsystem (i.e. immediate social settings including family, school, work), mesosystem (interactions between microsystems), exosystem (wider social systems such as neighbourhoods and community, which impact on microsystems), macrosystem (educational, social, economic and political systems) and ecosystem (the biosphere upon which all other levels depend) (Lomas, 2015). Ecological models provide a useful framework within which to work towards an interdisciplinary and even, transdisciplinary science of wellbeing, and all chapters—more or less—adopt such a view. We now provide a brief overview of the individual perspectives outlined in the preceding chapters and reflect on these in relation to the wider literature base. We focus initially on recent developments in the field of positive psychology, highlighting a move beyond the individual to a focus on complex societal issues. We then turn our attention to psychological interventions including positive psychotherapy and acceptance and commitment therapy which further demonstrate how psychological science is beginning to impact on the systems in which people are embedded. We then discuss two chapters that reinforce the relationships between human wellbeing, nature and its conservation, before shifting our focus from psychological science to public health, economics and sociopolitical systems.

Shifting the Landscape of Wellbeing Science

The first two chapters document the expansion of the field of positive psychology from a field emerging in the early 2000s that refocused attention on the science of positive emotions (Seligman et al., 2005; Seligman & Csikszentmihalyi, 2000) to a modern and boundary-pushing academic effort that applies insights to wide-ranging existential issues including suffering, poverty and climate justice. Paul Wong has been instrumental to the developments in positive psychology (as described in Wong, 2019). He and colleagues characterise these developments as representative of the emergent field of existential positive psychology, which focuses on integrating the dark (negative psychology) and bright (positive psychology) sides of life. This development is important because it provides a counter to common criticisms of positive psychology, focusing on the uncomfortable links between happiness, individualism and the modern consumer society (Carlisle et al., 2009; Davies, 2015; Ehrenreich, 2010; Purser, 2019). Positive psychology has been dominated by influential theories such as Seligman's PERMA model (Seligman, 2012), which focuses on positive emotions, engagement positive relationships, meaning and accomplishments. This model has helped to capture the key components of wellbeing, yet it is also challenged in the face of human suffering and existential threats to humanity (e.g. the climate emergency), as it has been unclear how traditional models are relevant to such experience. Accordingly, the authors reframe Seligman's PERMA model according to the perspective of existential positive psychology, emphasising positivity through embracing the dark side of life (e.g. existential gratitude) (Jans-Beken & Wong, 2019), engagement through painful discipline and avoiding distractions, strong relationships including forgiveness, finding meaning through suffering (e.g. tragic optimism) (Leung, 2019; Mead, Fisher, Tree, et al., 2021), and achievement following failure. This approach has considerable value, especially against the background of chronic disease burden (http://ihmeuw.org/5nrp) (Kemp et al., 2022), loneliness (Hawkley & Cacioppo, 2010) and the unfolding climate catastrophe (Ripple et al., 2021).

The complementary chapter by Lambert and colleagues (Chap. 3) presents three concrete examples of how positive psychology has begun to address systemic issues in society, including the promotion of non-Western views of wellbeing, egalitarianism and climate action. Historically, positive psychology and psychological science more generally, has been characterised—and criticised for—a focus on WEIRD samples, characterised as Western, educated, industrialised, rich and democratic (Hendriks et al., 2019). However, developments in the field, also described in the chapter by Wong and colleagues (Chap. 2), have begun to move beyond Western culture and promote non-Western views of wellbeing. In a related paper published elsewhere (Lambert et al., 2020), a more global view of wellbeing has been described, which encompasses a focus on social wellbeing, culture, nature and governance. Events such as the COVID-19 pandemic, the war in Ukraine and the unfolding climate catastrophe have further highlighted the interconnections and impacts of systems on individuals, redirecting attention to societal inequities and inequalities and reinforcing the need for more approach-oriented strategies to foster intergroup wellbeing (e.g. Atkins et al., 2019). Lambert's chapter (Chap. 3) also emphasises the principle of egalitarianism, illustrating a role for positive psychology in promoting prosocial behaviour and authentic allyship to support intergroup interactions. The authors end their chapter with a focus on the need for climate action and ecological approaches to wellbeing, a topic also tackled in chapters from Pritchard and Richardson (Chap. 6) as well as that from Corral-Verdugo and colleagues (Chap. 7). Lambert and colleagues argue that individual choices, collective behaviour and government action are needed to slow climate change and to live with it. We now turn our attention to how psychological interventions are beginning to be applied at multiple levels of scale to promote societal wellbeing after which we return to issues relating to the emergent field of 'climate psychology'.

Applying Psychological Interventions at Higher Levels of Scale

While psychological interventions such as positive psychotherapy and acceptance and acceptance and commitment therapy (ACT) have traditionally targeted the individual, researchers have begun to apply core principles at higher levels of scale to facilitate new opportunities for the promotion of individual, collective and planetary wellbeing. Application of these interventions are consistent with developments in the field, not only in positive psychology (Kern et al., 2020; Lomas, 2022; Lomas et al., 2021), but also in contextual behavioural science, the discipline within which ACT has been developed. While these interventions have been developed within their own respective fields, there is much complementarity, and recent work has sought unify these ideas within a multi-dimensional and multi-levelled evolutionary framework (Atkins et al., 2019; Ciarrochi et al., 2022). The chapters by Gibbs and colleagues (Chap. 4), and that by Edwards (Chap. 5) provide an example and context for how psychological interventions are being applied at multiple levels of scale.

The chapter by Gibbs and colleagues (Chap. 4) is focused on the healthcare sector and reflects on the challenges associated with the burden of chronic disease. The challenge here is that healthcare services are guided by an 'acute medical model' that is concentrated on the management of acute rather than chronic conditions, despite epidemiological evidence showing that chronic disease now outstrips that of acute conditions (http://ihmeuw.org/5nrp), a phenomenon associated with the epidemiological transition (Kemp et al., 2022; Smith, 2015). The authors describe how they have drawn on social ecological models, systems theory and wellbeing science to develop their own theoretical model of wellbeing (the GENIAL model) (Fisher et al., 2020; Kemp et al., 2017; Kemp & Fisher, 2022; Mead, Fisher, & Kemp, 2021), which has been used to reimagine opportunities for building health in communities, laying strong foundations for a more sustainable healthcare sector. Wellbeing is conceptualised at multiple levels including the individual, community and environment, while systems theory (Kern et al., 2020) provides a useful context within which these different levels and domains may be integrated. The authors describe how they have employed partnership and collaborative working across typically isolated organisations (e.g. academia, healthcare and community-based organisations) to overcome entrenched barriers (e.g. funding and limited numbers of staff working in a community brain injury service) to realise a common goal of building wellbeing in acquired brain injury. This approach to partnership working has wide applicability and is consistent with efforts to promote co-operative working within and across groups in the following chapter by Edwards (Chap. 5).

Societal challenges such as gender equality, persistent poverty and climate change are examples of complex problems that lack a clearly defined solution. Such challenges may be described as a 'wicked problem' as they trigger divergent views about the nature of the problem and responsibilities for addressing it (Head, 2022; Rittel & Webber, 1973). The question then arises how we might move beyond complexity, debate and inaction toward potential solutions that meaningfully impact those who most need them. The chapter by Edwards presents one such opportunity involving the promotion of cooperation within and across groups (Chap. 5). Edwards describes how acceptance and commitment therapy (ACT)—a psychological approach that supports individuals to identify and engage with meaningful intrinsic values—is now being combined with evolutionary principles within a 'Prosocial' framework (Atkins et al., 2019) to support individuals and groups to identify what is important to them and what they would like to work towards. This approach is argued to support selection of values that are adaptive and benefit the individual and group in much the same way as biological (Darwinian) evolution supports selection of genes, while avoiding critiques of 'social Darwinism'. This is achieved through what Elizabeth Ostrom has described as core design principles (CPDs) (Ostrom, 1990), which involve consciously focusing on equity, fairness, reward, self-determination, effective monitoring, conflict resolution and cooperation across groups. Edwards contrasts this approach with the dominant theoretical and ideological framework of 'neoliberalism', an extreme form of contemporary capitalism characterised by privatisation, limited public expenditure and the promotion of individual responsibility (Barnett & Bagshaw, 2020). It remains to be seen whether the 'Prosocial' approach will be able to effectively counter the harsh realities of modern competitive, individualistic and lonely 'neoliberal' societies (Becker et al., 2021; Morgan et al., 2022). Or, as Neilson argues in the final chapter (Chap. 10), is a complete overhaul of how societies function required? Neoliberal countries have been shown to perform worse in addressing the climate crisis (Fremstad & Paul, 2022), a topic that we turn to next.

Reflecting on the Role of Nature for the Construct of Wellbeing

Despite much discussion and commitment to addressing the climate crisis by governments in power, we remain on a 'business as usual' trajectory with major implications for human survival (Ripple et al., 2017, 2019, 2021). Young people not only report feeling threatened by the climate crisis, but also a sense of being unheard, betrayed and abandoned by older generations (Hickman et al., 2021; Morgan et al., 2022). In 2015, the Sustainable Development Goals (SDGs) provided a comprehensive plan for a world in which the planet and people would flourish by 2030, yet progress has been disappointing (Baum et al., 2022). There is a sense that we have lacked the inner capacity for managing an increasingly complex world, highlighting an important role for psychological science (https://www.innerdevelopmentgoals.org/) and the emergent field of climate psychology (https://www.climatepsychologyalliance.org/). The psychological roots of the climate crisis (Turp & Weintrobe, 2022; Weintrobe, 2021), including an inner exceptionalism and a culture of uncare, highlight a need for transformative action across multiple levels of scale, a common theme of our book.

The chapter by Pritchard and Richardson (Chap. 6) presents a call to arms for transformative change requiring a new relationship with nature involving a psychological connectedness to the natural environment. Such connection encompasses aspects of cognition and affect. For instance, higher levels of nature connectedness are associated with improvements in cognitive function including attention, memory, executive function, learning and creative potential, while impacts on affect encompass positive mood, improvements to psychological wellbeing and restorativeness (Capaldi et al., 2015; Kasap et al., 2021; Richardson et al., 2016). The authors further highlight that nature connectedness is associated with feelings of awe and self-transcendent experiences, reinforcing the relevance of nature to the promotion of existential positive psychology, described in the chapter by Wong and colleagues (Chap. 2). Unfortunately, humanity has become disconnected from nature—a phenomenon described as 'nature deficit disorder' (Louv, 2008)—resulting from a sedentary lifestyle, computer-based, desk-bound ways of working, and more time spent indoors. This disconnectedness has led to what Pritchard and Richardson describe as a 'narrow-minded and convenient placidity' that is contributing to the worsening climate crisis and loss in biodiversity. Some suggestions for how we might rekindle our relationship with nature including for example, appreciating nature's beauty, nature-conservation and care, as well as the need for individual and collective action to cultivate connections to nature. Critically the authors move beyond nature connection at an individual level to an improved collective relationship that seeks to understand and alter the very value systems and worldviews that are contributing to the ecological crisis.

This work complements the chapter from Corral-Verdugo and colleagues (Chap. 7), which focuses on 'positive environments', defined as sustainable environments that facilitate the capacity to flourish, and the experience of health and wellbeing. The authors illustrate the interrelationships between resources, enablers and challenges that support the promotion of wellbeing. Environmental resources span the tangible (e.g. settings for fitness and psychological restoration) and intangible (e.g. social elements to encourage interaction and inclusion). Positive environments also provide pro-environmental enablers, including for example, environmental education programs and community programs to promote conservation. Such enablers then facilitate pro-social and pro-environmental behaviours that help to promote individual and group actions needed to overcome environmental challenges such as poverty, environmental degradation and climate change. The authors draw on *Dynamic Equilibrium Theory* (Kloep et al., 2009) to highlight a need for balance across resources, enablers and challenges to support the experience of wellbeing in "a recursive chain of positive interdependencies". When the three elements are out of balance, it can lead to unfavourable outcomes. For instance, an unsustainable environment will lead to resource depletion, while an environment without challenges will lead to fewer opportunities for self-transcendence. We now shift our attention to complementary perspectives from public health, economic theory and socio-political systems.

Reflecting on Systems and Their Transformation to Promote Societal Wellbeing

The following chapters present perspectives from colleagues in other disciplines who have begun to think deeply about the construct of wellbeing within those disciplines. The chapter by Edmonds (Chap. 8) places the promotion of wellbeing within the context of public health, focused on Wales, an emergent wellbeing economy (https://weall.org/wego).

Edmonds defines wellbeing holistically highlighting aspirations for complete physical, mental and social wellbeing across multiple levels and settings within which people live, work and learn. She further argues that wellbeing researchers face a major challenge when seeking to promote wellbeing at multiple levels of scale. She writes:

> "The definitions and frameworks of health and wellbeing… will ultimately shape the design of policies/programmes and what type of responses (if any) are developed… If wellbeing is only viewed at an individual level, opportunities for investment in building community assets and resilience may not be within scope."

She describes how the Welsh Government has begun to measure wellbeing at multiple levels of scale, using a set of 50 indicators that include measures of individual, community, cultural, economic and planetary wellbeing, an aim embedded into the Wellbeing of Future Generations (Wales) Act (2015). She then provides examples of how public health tools such as the Health Impact Assessment (HIA) and Mental Wellbeing Impact Assessment (MWIA) have helped to develop shared understandings of wellbeing and to shift systems to focus on the determinants of wellbeing. These tools aid the measurement of the impacts of wellbeing interventions across multiple levels of scale, which may then feed into wellbeing public policy (Fabian & Pykett, 2021). Ultimately the goal for public health practitioners is to successfully influence the distribution of resource allocation to promote wellbeing at scale, although their efforts are constrained by economic assumptions, the focus of the chapter by Webber and Page (Chap. 9).

Mainstream neoclassical economic theory implicitly assumes that individuals seek to maximise levels of wellbeing, in much the same that firms seek to maximise their profits. Webber and Page suggest that this assumption ignores the impacts of social networks, relative social position and poverty. The point here is that individual wellbeing is dependent on other social-contextual factors that lie beyond the control of the individual themselves, questioning the assumption that individuals have sufficient agency to act in their own self-interest. The authors contrast the neoclassical economic approach with a more holistic perspective that recognises that the individual experience of wellbeing will be impacted on by factors within and beyond individual control, consistent with ecological models. While researchers in the field are beginning to take these wider societal impacts on wellbeing into consideration, challenges remain. For instance, the authors present a paradox relating to the profitability of drug companies focused on the cure of 'wellbeing problems'. On the one hand business profitability would suggest that the economy is headed in the right direction, yet on the other, questions should be asked what is contributing to the demand in the first place. The implication here is that the system in which individuals are embedded are contributing to the rise of existential anxieties and mental ill-health, and there is recent evidence for such (Becker et al., 2021). This issue is picked up in the final chapter by Neilson (Chap. 10), who casts a critical eye on the neoliberal model of development, emphasising that present day social and material circumstances in the Western world are at odds with what it needs to flourish. According to Neilson, an alternative must be grounded in universal friendship, creative cooperation and material security, underpinned by local self-sufficiency and conscious, cooperative nationalism, an agenda that may be facilitated by previously described efforts to promote prosociality (Atkins et al., 2019) at multiple levels of scale. Neilson argues that cooperative nationalism must be supported by a reformed United Nations that is focused on global solidarity according to universal conditions for wellbeing and ecological sustainability.

In summary, our book has provided a glimpse into how wellbeing scholars have begun to respond to some of the major societal challenges facing humanity including the inevitability of human suffering driven by persistent poverty, inequity and climate breakdown. Wellbeing science is 'a broad church', with developments taking place in a wide range of fields focused on the individual through to public health as well as economic and political systems, and all book chapter authors have highlighted—more or less—how researchers in their fields are beginning to reach beyond their individual disciplines to conceptualise and apply wellbeing science in theory and practice. There is tremendous scope for the development and application of ideas presented herein, as well as in the monitoring of the impacts that such applications are having at multiple levels. While there is much to do, we believe that we have laid a strong grounding for the field of wellbeing science to move forward in such a way that seeks to extend beyond the boundaries of single disciplines and competing therapeutic approaches, with a focused eye on tackling major societal challenges, guided by complementary theories and insights emerging in the field. It is an exciting time for wellbeing researchers, and we hope that this book will be a source of inspiration for many years to come.

References

Atkins, P. W. B., Wilson, D. S., & Hayes, S. C. (2019). *Prosocial: Using evolutionary science to build productive, equitable, and collaborative groups.* New Harbinger Publications.

Barnett, P., & Bagshaw, P. (2020). Neoliberalism: What it is, how it affects health and what to do about it. *The New Zealand Medical Journal, 133*(1512), 76–84.

Baum, F., Paremoer, L., Flavel, J., Musolino, C., & Labonte, R. (2022). Can the world become a place where the planet and all people flourish after the pandemic? *British Medical Journal, 377*, e067872. https://doi.org/10.1136/bmj-2021-067872

Becker, J. C., Hartwich, L., & Haslam, S. A. (2021). Neoliberalism can reduce well-being by promoting a sense of social disconnection, competition, and loneliness. *British Journal of Social Psychology, 60*(3), 947–965. https://doi.org/10.1111/bjso.12438

Bronfenbrenner, U. (1977). Toward an experimental ecology of human development. *American Psychologist, 32*(7), 513–531. https://doi.org/10.1037/0003-066x.32.7.513

Capaldi, C. A., Passmore, H.-A., Nisbet, E. K., Zelenski, J. M., & Dopko, R. L. (2015). Flourishing in nature: A review of the benefits of connecting with nature and its application as a wellbeing intervention. *International Journal of Wellbeing, 5, 1.*

Carlisle, S., Henderson, G., & Hanlon, P. W. (2009). 'Wellbeing': A collateral casualty of modernity? *Social Science & Medicine, 69*(10), 1556–1560. https://doi.org/10.1016/j.socscimed.2009.08.029

Ciarrochi, J., Hayes, S. C., Oades, L. G., & Hofmann, S. G. (2022). Toward a unified framework for positive psychology interventions: Evidence-based processes of change in coaching, prevention, and training. *Frontiers in Psychology, 12*, 809362. https://doi.org/10.3389/fpsyg.2021.809362

Davies, W. (2015). *The happiness industry.* Verso Books.

Ehrenreich, B. (2010). *Smile or die: How positive thinking fooled America and the world.* Granta.

Fabian, M., & Pykett, J. (2021). Be happy: Navigating normative issues in behavioral and well-being public policy. *Perspectives on Psychological Science, 17*, 174569162098439. https://doi.org/10.1177/1745691620984395

Fisher, Z., Galloghly, E., Boglo, E., Gracey, F., & Kemp, A. (2020). Emotion, wellbeing and the neurological disorders. In *Reference Module in Neuroscience and Biobehavioral Psychology.* Elsevier. https://doi.org/10.1016/b978-0-12-819641-0.00013-x.

Fremstad, A., & Paul, M. (2022). Neoliberalism and climate change: How the free-market myth has prevented climate action. *Ecological Economics, 197*, 107353. https://doi.org/10.1016/j.ecolecon.2022.107353

Hawkley, L. C., & Cacioppo, J. T. (2010). Loneliness matters: A theoretical and empirical review of consequences and mechanisms. *Annals of Behavioral Medicine, 40*(2), 218 227. https://doi.org/10.1007/s12160-010-9210-8

Head, B. W. (2022). *Wicked problems in public policy, understanding and responding to complex.* Challenges. https://doi.org/10.1007/978-3-030-94580-0

Hendriks, T., Warren, M. A., Schotanus-Dijkstra, M., Hassankhan, A., Graafsma, T., Bohlmeijer, E., & de Jong, J. (2019). How WEIRD are positive psychology interventions? A bibliometric analysis of randomized controlled trials on the science of well-being. *The Journal of Positive Psychology, 14*, 1–13. https://doi.org/10.1080/17439760.2018.1484941

Hickman, C., Marks, E., Pihkala, P., Clayton, S., Lewandowski, R. E., Mayall, E. E., Wray, B., Mellor, C., & van Susteren, L. (2021). Climate anxiety in children and young people and their beliefs about government responses to climate change: A global survey. *The Lancet Planetary Health, 5*(12), e863–e873. https://doi.org/10.1016/s2542-5196(21)00278-3

Jans-Beken, L., & Wong, P. T. P. (2019). Development and preliminary validation of the existential gratitude scale (EGS). *Counselling Psychology Quarterly, 34*, 1–15. https://doi.org/10.1080/09515070.2019.1656054

Kasap, E. Z., Ağzıtemiz, F., & Ünal, G. (2021). Cognitive, mental and social benefits of interacting with nature: A systematic review. *Journal of Happiness and Health, 1*(1), 16–27.

Kemp, A. H., & Fisher, Z. (2022). Wellbeing, whole health and societal transformation: Theoretical insights and practical applications. *Global Advances in Health and Medicine, 11*, 21649561211073076. https://doi.org/10.1177/21649561211073077

Kemp, A. H., Arias, J. A., & Fisher, Z. (2017). Social ties, health and wellbeing: A literature review and model. In A. Ibáñez, L. Sedeño, & A. García (Eds.), *Neuroscience and Social Science* (Vol. 59, pp. 397–427). Springer International Publishing. https://doi.org/10.1007/978-3-319-68421-5_17

Kemp, A. H., Tree, J., Gracey, F., & Fisher, Z. (2022). Editorial: Improving wellbeing in patients with chronic conditions: Theory, evidence, and opportunities. *Frontiers in Psychology, 13*, 868810. https://doi.org/10.3389/fpsyg.2022.868810

Kern, M. L., Williams, P., Spong, C., Colla, R., Sharma, K., Downie, A., Taylor, J. A., Sharp, S., Siokou, C., & Oades, L. G. (2020). Systems informed positive psychology. *The Journal of Positive Psychology, 15*(6), 705–715. https://doi.org/10.1080/17439760.2019.1639799

Kloep, M., Hendry, L. B., & Saunders, D. (2009). A new perspective on human development. *Conference of the International Journal of Arts and Sciences, 1*(6), 332–343.

Lambert, L., Lomas, T., van de Weijer, M. P., Passmore, H. A., Joshanloo, M., Harter, J., Ishikawa, Y., Lai, A., Kitagawa, T., Chen, D., Kawakami, T., Miyata,

H., & Diener, E. (2020). Towards a greater global understanding of wellbeing: A proposal for a more inclusive measure. *International Journal of Wellbeing, 10*(2), 1–18. https://doi.org/10.5502/ijw.v10i2.1037

Leung, M. M. (2019). Tragic optimism: An integrative meaning-centred approach to trauma treatment. *Counselling Psychology Quarterly, 32*(3–4), 1–19. https://doi.org/10.1080/09515070.2019.1633497

Lomas, T. (2015). Positive social psychology: A multilevel inquiry into sociocultural well-being initiatives. *Psychology Public Policy and Law, 21*(3), 338–347. https://doi.org/10.1037/law0000051

Lomas, T. (2022). Making waves in the great ocean: A historical perspective on the emergence and evolution of wellbeing scholarship. *The Journal of Positive Psychology, 17*(2), 257–270. https://doi.org/10.1080/17439760.2021.2016900

Lomas, T., Hefferon, K., & Ivtzan, I. (2015). The LIFE model: A meta-theoretical conceptual map for applied positive psychology. *Journal of Happiness Studies, 16*(5), 1347–1364. https://doi.org/10.1007/s10902-014-9563-y

Lomas, T., Waters, L., Williams, P., Oades, L. G., & Kern, M. L. (2021). Third wave positive psychology: Broadening towards complexity. *The Journal of Positive Psychology, 16*, 1–15. https://doi.org/10.1080/17439760.2020.1805501

Louv, R. (2008). *Last child in the woods: Saving our children from nature-deficit disorder*. Algonquin Books. Updated and Expanded edition.

Mead, J., Fisher, Z., & Kemp, A. H. (2021). Moving beyond disciplinary silos towards a transdisciplinary model of wellbeing: An invited review. *Frontiers in Psychology, 12*, 642093. https://doi.org/10.3389/fpsyg.2021.642093

Mead, J. P., Fisher, Z., Tree, J. J., Wong, P. T. P., & Kemp, A. H. (2021). Protectors of wellbeing during the COVID-19 pandemic: Key roles for gratitude and tragic optimism in a UK-based cohort. *Frontiers in Psychology, 12*, 647951. https://doi.org/10.3389/fpsyg.2021.647951

Morgan, G., Barnwell, G., Johnstone, L., Shukla, K., & Mitchell, A. (2022). The power threat meaning framework and the climate and ecological crises. *Psychology in Society, 63*, 83–109.

Ostrom, E. (1990). *Governing the commons: The evolution of institutions for collective action*. Cambridge University Press.

Purser, R. (2019). *McMindfulness: How mindfulness became the new capitalist spirituality*. Repeater Books.

Richardson, M., McEwan, K., Maratos, F., & Sheffield, D. (2016). Joy and calm: How an evolutionary functional model of affect regulation informs positive emotions in nature. *Evolutionary Psychological Science, 2*(4), 308–320. https://doi.org/10.1007/s40806-016-0065-5

Ripple, W. J., Wolf, C., Newsome, T. M., Galetti, M., Alamgir, M., Crist, E., Mahmoud, M. I., & Laurance, W. F. (2017). World scientists' warning to humanity: A second notice. *Bioscience, 67*(12), 1026–1028. https://doi.org/10.1093/biosci/bix125

Ripple, W. J., Wolf, C., Newsome, T. M., Barnard, P., & Moomaw, W. R. (2019). World scientists' warning of a climate emergency. *Bioscience, 70*, 8–12. https://doi.org/10.1093/biosci/biz088

Ripple, W. J., Wolf, C., Newsome, T. M., Gregg, J. W., Lenton, T. M., Palomo, I., Eikelboom, J. A. J., Law, B. E., Huq, S., Duffy, P. B., & Rockström, J. (2021). World scientists' warning of a climate emergency 2021. *Bioscience, 71*, 894. https://doi.org/10.1093/biosci/biab079

Rittel, H. W. J., & Webber, M. M. (1973). Dilemmas in a general theory of planning. *Policy Sciences, 4*(2), 155–169. https://doi.org/10.1007/bf01405730

Seligman, M. (2012). *Flourish: A visionary new understanding of happiness and well-being*. Simon and Schuster.

Seligman, M. E. P., & Csikszentmihalyi, M. (2000). Positive psychology: An introduction. *American Psychologist, 55*(1), 5–14. https://doi.org/10.1037//0003-066x.55.1.5

Seligman, M. E. P., Steen, T. A., Park, N., & Peterson, C. (2005). Positive Psychology Progress: Empirical validation of interventions. *American Psychologist, 60*(5), 410–421. https://doi.org/10.1037/0003-066x.60.5.410

Smith, J. N. (2015). *Epic measures: One doctor. Seven billion patients.* HarperCollins Publishers.

Turp, M., & Weintrobe, S. (2022). Exploring the psychological roots of the climate crisis. *Psychodynamic Practice, 1–7*, 1. https://doi.org/10.1080/14753634.2022.2071326

Weintrobe, S. (2021). *Psychological roots of the climate crisis: Neoliberal exceptionalism and the culture of uncare.* Bloomsbury Publishing. https://books.google.co.uk/books/about/Psychological_Roots_of_the_Climate_Crisi.html?id=ez0ZEAAAQBAJ&source=kp_book_description&redir_esc=y.

Wong, P. T. P. (2019). Second wave positive psychology's (PP 2.0) contribution to counselling psychology. *Counselling Psychology Quarterly, 32*, 1–10. https://doi.org/10.1080/09515070.2019.1671320

Correction to: Beyond Us: Building Collective Wellbeing

Louise Lambert, Meg A. Warren, Rachael J. Waldrop, Sabrina Al Subaiei, Alecia Eubanks, and Jen L. Fisher

Correction to:

Chapter 3 in: A. H. Kemp, D. J. Edwards (eds.),
***Broadening the Scope of Wellbeing Science*,**
https://doi.org/10.1007/978-3-031-18329-4_3

The original version of this chapter was published with an incorrect spelling of the author's name as Rachel J. Waldrop. The correct name Rachael J. Waldrop has been updated.

The updated original version for this chapter can be found at
https://doi.org/10.1007/978-3-031-18329-4_3

A. H. Kemp, D. J. Edwards (eds.), *Broadening the Scope of Wellbeing Science*, https://doi.org/10.1007/978-3-031-18329-4_12

Index[1]

[1] Note: Page numbers followed by 'n' refer to notes.

A. H. Kemp, D. J. Edwards (eds.), *Broadening the Scope of Wellbeing Science*, https://doi.org/10.1007/978-3-031-18329-4

The manufacturer's authorised representative in the EU is Springer Nature Customer Service Centre GmbH, Europaplatz 3, 69115 Heidelberg, Germany. If you have any concerns regarding our products, please contact ProductSafety@springernature.com

Printed and bound by CPI Group (UK) Ltd, Croydon, CR0 4YY
28/09/2026
02224074-0001